Multilingual Dictionary of Nuclear Reactor Physics and Engineering

Multilingual Dictionary of Nuclear Reactor Physics and Engineering

Henryk Anglart

CRC Press
Taylor & Francis Group
Boca Raton London New York

CRC Press is an imprint of the
Taylor & Francis Group, an **informa** business

First edition published 2021
by CRC Press
6000 Broken Sound Parkway NW, Suite 300, Boca Raton, FL 33487-2742

and by CRC Press
2 Park Square, Milton Park, Abingdon, Oxon, OX14 4RN

© 2021 Taylor & Francis Group, LLC

CRC Press is an imprint of Taylor & Francis Group, LLC

Library of Congress Cataloging-in-Publication Data

Names: Anglart, Henryk, author.
Title: Multilingual dictionary of nuclear reactor physics and engineering /
Henryk Anglart.
Description: First edition. | Boca Raton : CRC Press, 2020. | Includes
index. | In English, with French, German, Swedish, and Polish
translations.
Identifiers: LCCN 2020010210 | ISBN 9780367470814 (hardback) | ISBN
9781003037019 (ebook)
Subjects: LCSH: Nuclear reactors--Dictionaries--Polyglot. | Nuclear
physics--Dictionaries--Polyglot. | Nuclear
engineering--Dictionaries--Polyglot.
Classification: LCC TK9202 .A56 2020 | DDC 621.48/303--dc23
LC record available at https://lccn.loc.gov/2020010210

ISBN: 978-0-367-47081-4 (hbk)
ISBN: 978-1-003-03701-9 (ebk)

Typeset in Computer Modern font
by KnowledgeWorks Global Ltd

To my Family
Ewa, Dorota, and Marcin

Preface

The purpose of this dictionary is to provide concise, reliable, and useful definitions of terms used in modern nuclear reactor physics and engineering. The conciseness has been achieved by a simplification and compression of definitions, consisting usually of one, or a few, semicolon-separated clauses. For high reliability, many credible sources of information have been used, such as peer-reviewed multilingual technical dictionaries, books, and journal papers. The terms have been selected with their practical usefulness in mind, assuming that users are students, researchers, or practitioners of nuclear reactor physics and engineering or a related discipline.

Even though the dictionary is intentionally kept compact, every effort was made to help the user to easily grasp all the information contained in an entry. The main entries, which contain all explanatory text and relevant references, are well spaced and begin with a single headword or head expression printed in a bold text. The headword is followed by an abbreviation of the sub-domain in nuclear engineering to which the headword belongs. The definition part of the main entry begins after a bullet symbol (•) and is followed by an arrow symbol ($\mapsto$). The arrow indicates additional related headwords separated by semicolons, which are suggested for further reading. Equivalents of the headword in other languages (German, French, Polish, and Swedish) are provided at the end of the entry.

The dictionary can be used for various purposes, such as searching for explanations of new terms, and translating the terms into other languages; however, it is also a helpful reference tool for all its users. The section "Guide to the Dictionary" outlines the principles upon which each aspect of the dictionary has been planned. A list of the most commonly used technical abbreviations is contained in the section "Abbreviations". Most of the listed abbreviations are further explained in the dictionary, and the corresponding page number is provided.

The dictionary contains approximately 1500 English terms, with their equivalents in four additional languages. A dictionary of this scope, besides the author, owes much to many people, who, over the past half-century or so, have developed the new and fascinating discipline of nuclear engineering, and who have done the hard work to develop the definitions and the vocabulary used in this book. In this regard, I would like to acknowledge the resources provided by Sweden's National Term Bank (Rikstermbanken, www.rikstermbanken.se) for the Swedish terms and their German and French equivalents, the *English-*

Polish Dictionary of Science and Technology for the Polish translations, and the *IAEA Safety Glossary* for the modern definitions of terms used in nuclear power safety. The feedback and support from many people, in particular students and faculties at the KTH Royal Institute of Technology and the Warsaw University of Technology, as well as the staff at Taylor & Francis, helped me to finalize this dictionary, and I am very grateful to all of them.

Stockholm
January 2020
Henryk Anglart

Guide to the Dictionary

The dictionary can be used in multiple ways, depending on a particular need of the user. The most straightforward usage is to search a meaning of an English term or expression. The English terms have been arranged in a strictly alphabetic order, that follows the letter-by- letter system. If a term has one or more synonyms, all synonyms are listed in a note given within round brackets and placed at the beginning of the body text (see section "The Main Entry" for a detailed description of the entry structure). The synonyms are listed in the dictionary as separate entries, but contain a cross-reference to the main entry only. In that way, an explanation for all terms occurs only once, but it can be easily located in the dictionary through any of the known synonyms.

Another usage of the dictionary can be to search an equivalent of an English term in any of the following languages: German, French, Polish, or Swedish. These multilingual equivalents are provided at the end of the entry, following the cross-references to related entries. When a term of interest is known in any of the above-mentioned languages, the English equivalent can be found in the corresponding language index. In that way, the full description of the term, and also other language equivalents, can be accessed.

THE MAIN ENTRY

The example below shows the main entry for the headword **condensation pool**. The headword belongs to sub-domain "reactor components and systems", as indicated by the abbreviation *rcs*. The list of abbreviations for all sub-domains included in the dictionary is given in section "Abbreviations of Sub-Domains".

The body text that follows the headword can be preceded by text in parentheses, ended with a colon, and called a "note". The purpose of the note is to provide some auxiliary information about the headword such as a limitation of the application, synonymous expressions, a typical notation used in relation to the headword, and similar. The headword is frequently repeated in the body text, and to limit the space usage, only the headword's initials, for better visibility printed in italics, are used. Cross-references to related entries are placed after the body text and are preceded by a symbol $\mapsto$ for better visibility.

Headwords that have identical spellings but different meanings or different multilingual equivalents are inserted separately and numbered with an upper

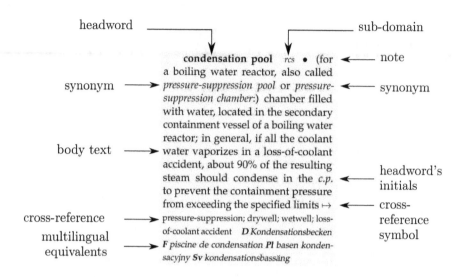

index, for example **enrichment**[1] and **enrichment**[2]. For headwords with different meanings but identical spellings in all languages, a single entry is used with Arabic numerals **1.**, **2.**, etc.

Some entries contain sub-headwords, which for better visibility are printed in italics. The sub-headwords are alphabetically listed in the dictionary together with all other headwords. For example, the entry **enrichment**[1] *nf* • **1.** ... *enrichment techniques* ... contains a sub-headword *enrichment techniques*, and this phrase is separately listed in the dictionary as **enrichment techniques** ↦*enrichment*[1]. In this way, the reader searching for "enrichment techniques" will be re-directed to the headword **enrichment**[1].

THE CROSS-REFERENCE ENTRY

The cross-reference entry contains a cross-reference only and is used to facilitate the finding of the relevant main entry. This feature is particularly useful to resolve the following situations:

- to identify a symbol or an abbreviation, e.g., **A** ↦*mass number,*

- to define a synonym, e.g., **bulk boiling** ↦*saturated boiling,*

- to provide a reference to the main entry that contains the explanation of the headword, e.g., **actual quality** ↦*steam quality.*

ABBREVIATIONS OF SUB-DOMAINS

bph	basic physics
gnt	general nuclear technology
mat	materials and material properties
mt	measuring technique
mth	mathematics
nap	nuclear and atomic processes
nch	nuclear chemistry
nf	nuclear fuel
rcs	reactor components and systems
rd	radiation
rdp	radiation protection
rdy	radioactivity
roc	reactor operation and control
rph	reactor physics
rs	reactor safety
rty	reactor type
sfg	safeguards (of nuclear material)
th	thermal engineering
wst	waste
xr	cross sections and resonances

ABBREVIATIONS OF LANGUAGES

D	German
E	English
F	French
(GB)	British English
Pl	Polish
Sv	Swedish
(US)	American English

OTHER ABBREVIATIONS AND SYMBOLS

d	day
g	terrestrial acceleration
h	hour
ky	thousand years
L	liter
m	month
My	million years
pl	plural
y	year
$\mapsto$	see related headwords

Abbreviations

Some commonly used abbreviations in nuclear reactor physics and engineering are listed below. For abbreviations that are further addressed in the dictionary, the corresponding page numbers are provided.

ABWR	*Advanced Boiling Water Reactor*
ACR	*Advanced CANDU Reactor (p. 4)*
ADS	*Accelerator Driven System (p. 3)*
ADS	*Automatic Depressurization System*
AECL	*Atomic Energy of Canada Limited*
AGR	*Advanced Gas-cooled Reactor (p. 5)*
ALARA	*As Low as Reasonably Achievable (p. 6)*
AOA	*Axial Offset Anomaly (p. 38)*
AOO	*Anticipated Operational Occurrence*
ASN	*Autorité de Sûreté Nucléaire (nuclear regulatory authority in France)*
ATWS	*Anticipated Transient Without Scram (p. 8)*
BA	*Burnable Absorber (p. 21)*
BOC	*Beginning of Cycle (p. 14)*
BR	*Breeding Ratio (p. 20)*
BWR	*Boiling Water Reactor (p. 17)*
CANDU	*CANada Deuterium Uranium (reactor) (p. 23)*
CHF	*Critical Heat Flux (p. 36)*
CIM	*Conductivity Integral to Melt (p. 164)*
CIPS	*Crud Induced Power Shift (p. 38)*
CPR	*Critical Power Ratio (p. 37)*
CR	*Conversion Ratio (p. 33)*
CRUD	*Chalk River Unidentified Deposit (p. 38)*
CVCS	*Chemical and Volume Control System (p. 26)*
DBA	*Design Basis Accident (p. 43)*
DBE	*Design Basis Earthquake (p. 43)*
DNB	*Departure from Nucleate Boiling (p. 42)*
DNBR	*Departure from Nucleate Boiling Ratio (p. 42)*
EAL	*Emergency Action Level (p. 139)*
ECCS	*Emergency Core Cooling System (p. 53)*
EFPH	*Effective Full Power Hours (p. 51)*
ENDF	*Evaluated Nuclear Data File*
ENSREG	*European Nuclear Safety Regulators Group*

EOC	*End of Cycle (p. 54)*
EPR	*European Pressurized water Reactor (p. 56)*
EPZ	*Emergency Planning Zone (p. 54)*
ESBWR	*Economic Simplified Boiling Water Reactor*
FBR	*Fast Breeder Reactor*
FIFA	*Fissions per Initial Fissile Atom (p. 61)*
FIMA	*Fissions per Initial Metal Atom (p. 61)*
FSAR	*Final Safety Analysis Report (p. 145)*
GCFR	*Gas-Cooled Fast Breeder Reactor*
GCR	*Gas-Cooled Reactor (p. 69)*
GDCS	*Gravity-Driven Cooling System*
GFR	*Gas-cooled Fast Reactor (p. 69)*
HEU	*Highly Enriched Uranium (p. 75)*
HTGR	*High Temperature Gas-cooled Reactor (p. 75)*
HTR	*High Temperature Reactor*
HVAC	*Heating, Ventilation and Air-Conditioning*
IAEA	*International Atomic Energy Agency*
IASC	*Irradiation Assisted Stress Corrosion (p. 84)*
IASCC	*Irradiation Assisted Stress Corrosion Cracking*
ICFM	*In-Core Fuel Management (p. 79)*
ICRP	*International Commission on Radiological Protection (p. 83)*
IGSCC	*Intergrannular Stress Corrosion Cracking (p. 82)*
INES	*International Nuclear Event Scale (p. 83)*
KMP	*Key Measurement Point (p. 89)*
LEU	*Low Enriched Uranium (p. 94)*
LFR	*Lead-cooled Fast Reactor (p. 92)*
LOCA	*Loss of Coolant Accident (p. 94)*
LPRM	*Local Power Range Monitor (p. 94)*
LWR	*Light Water Reactor (p. 92)*
MBA	*Material Balance Area (p. 98)*
MBP	*Material Balance Period (p. 98)*
MCA	*Maximum Credible Accident (p. 99)*
MEU	*Medium Enriched Uranium (p. 100)*
MSR	*Molten Salt Reactor (p. 104)*
MUF	*Material Unaccounted For (p. 99)*
NEA	*Nuclear Energy Agency*
NRC	*Nuclear Regulatory Commission*
NSSS	*Nuclear Steam Supply System (p. 115)*
OBE	*Operating Basis Earthquake (p. 117)*
ONR	*Office for Nuclear Regulation*
PAA	*Polska Agencja Atomistyki (Polish National Atomic Energy Agency)*
PCCS	*Passive Containment Cooling System*
PCI	*Pellet-Clad Interaction (p. 120)*
PHWR	*Pressurized Heavy Water Reactor (p. 121)*
PIUS	*Process Inherent Ultimate Safety*
PRA	*Probabilistic Risk Analysis (p. 125)*

PRM	*Power Range Monitor (p. 123)*
PSA	*Probabilistic Safety Analysis (p. 125)*
PSAR	*Preliminary Safety Analysis Report (p. 145)*
PWR	*Pressurized Water Reactor (p. 125)*
RIA	*Reactivity-Induced Accident (p. 136)*
RIP	*Recirculation Internal Pump (p. 139)*
RPV	*Reactor Pressure Vessel (p. 138)*
SAR	*Safety Analysis Report (p. 145)*
SBWR	*Simplified Boiling Water Reactor*
SCWR	*Supercritical Water-cooled Reactor (p. 160)*
SFR	*Sodium-cooled Fast Reactor (p. 153)*
SSE	*Safe Shutdown Earthquake (p. 145)*
SSM	*Strålsäkerhetsmyndigheten (Swedish Radiation Safety Authority)*
TIP	*Travelling In-core Probe (p. 169)*
USNRC	*U.S. Nuclear Regulatory Commission*
VHTR	*Very-High-Temperature Reactor (p. 173)*
VFR	*Volume Reduction Factor (p. 174)*
WANO	*World Association of Nuclear Operators*
WNA	*World Nuclear Association*
WNU	*World Nuclear University*

Numerical Terms

1/v-detector *mt* • neutron detector in which the employed nuclear cross section is inversely proportional to the neutron speed (it follows the 1/v law) ↦ 1/v law *D* *1/v-Detektor* *F* *détecteur en 1/v* *Pl* *detektor 1/v* *Sv* *1/v-detektor*

1/v law *xr* • the statement that a certain neutron cross section is inversely proportional to the neutron velocity relative to the nucleus ↦ 1/v-detector *D* *1/v-Gesetz* *F* *loi en 1/v* *Pl* *prawo 1/v* *Sv* *1/v-lagen*

2200-meter-per-second flux density ↦ *conventional flux density*

Aa

A $\mapsto$ *mass number*

absolute temperature *th* • (also called *thermodynamic temperature:*) temperature measured from absolute zero and expressed in kelvins $\mapsto$ kelvin **D** *absolute Temperatur* **F** *température thermodynamique* **Pl** *temperatura absolutna* **Sv** *absolut temperatur*

absorbed dose *rdp* • measure of radiation energy absorbed by matter defined as

$$D \equiv \lim_{\Delta m \to 0} \frac{\Delta \bar{\epsilon}}{\Delta m}$$

where $\Delta \bar{\epsilon}$ is the mean energy imparted by ionizing radiation to matter in a volume element and Δm is the mass of matter in that volume element; the concept of *a.d.* is useful in radiation protection since energy imparted per unit mass is closely correlated with radiation hazard; traditional unit of *a.d.* was the rad, which is presently replaced with the standard SI unit gray (Gy) $\mapsto$ radiation protection; kerma; rad; gray[2] **D** *Energiedosis* **F** *dose absorbée* **Pl** *dawka pochłonięta* **Sv** *absorberad dos*

absorber *nap* • substance absorbing the energy of radiation $\mapsto$ absorption; neutron absorption; burnable absorber **D** *Absorber; Absorptionsmittel* **F** *absorbeur* **Pl** *absorbent* **Sv** *absorbator*

absorption *nap* • interception of radiant energy or sound waves $\mapsto$ neutron absorption; alpha radiation; beta radiation; gamma radiation **D** *Absorption* **F** *absorption* **Pl** *absorpcja; pochłanianie* **Sv** *absorption*

absorption control *roc* • reactor control through absorbers, whose properties, locations or amounts are changing $\mapsto$ reactor control[2]; absorber **D** *Steuerung durch Absorption* **F** *commande par absorption* **Pl** *sterowanie absorbentem* **Sv** *absorbatorstyrning*

abundance ratio *nf* • ratio of the number of atoms of two different isotopes of an element in a given sample $\mapsto$ isotope; isotopic abundance; isotope separation **D** *Isotopenhäufigkeitsverhältnis* **F** *rapport des teneurs (isotopiques)* **Pl** *stosunek zawartości izotopów* **Sv** *isotopkvot*

Ac $\mapsto$ *actinium*

accelerator-driven system *rty* • (previously referred to as energy amplifier, abbreviated ADS:) sub-critical nuclear fission reactor in which an accelerator is used to provide high-energy protons to bombard a heavy-metal target, such as lead, and to generate twenty to thirty neutrons per event through spallation; *a.-d.s.* does not rely on delayed neutrons for the control $\mapsto$ subcritical; accelerator; spallation **D** - **F** - **Pl** *system zasilany akceleratorem* **Sv** *acceleratordrivet system*

accident *rs* • any unintended event, including operating errors, equipment failures and other mishaps, the consequences or potential consequences of which are not negligible from the point of protection or safety ↦ nuclear accident; INES **D** *Störfall* **F** *accident* **Pl** *awaria* **Sv** *haveri*

accident management *rs* • planning and performing steps to mitigate consequences of an accident in a nuclear facility ↦ nuclear accident; emergency preparedness **D** *Störfall-Management* **F** *gestion de l'accident* **Pl** *zarządzanie awarią* **Sv** *haverihantering*

accident prevention *rs* • procedure to prevent an occurrence of operation disturbance that can lead to a core accident ↦ core accident; nuclear accident; protective system **D** *Störfall-Prevention* **F** *prévention des accident* **Pl** *zapobieganie awariom* **Sv** *haveriförebyggande åtgärd*

accountancy *sfg* • (related to safeguards of nuclear materials:) quantitative accounting of nuclear materials according to the national and international commitments ↦ nuclear material; safeguards (of nuclear materials) **D** *Buchführung des Kernmaterials* **F** *gestion* **Pl** *bilans materiałów rozszczepialnych* **Sv** *redovisning*

accumulator injection sys-tem ↦*accumulator system*

accumulator system *rcs* • passive emergency core-cooling subsystem containing two or more independent tanks containing cool borated water stored under nitrogen gas at a pressure of about 1.4 to 4.1 MPa; the tanks are connected through check valves to the reactor cold legs or sometimes directly to the reactor pressure vessel ↦ emergency core cooling system; passive system **D** *Akkumulatorsystem* **F** *circuit des accumulateurs* **Pl** *system hydroakumulatorów* **Sv** *ackumulatorsystem*

ACR *rty* • (acronym for *A*dvanced *C*ANDU *R*eactor:) generation III+ nuclear reactor designed by Atomic Energy of Canada Limited; it combines features of the existing CANDU pressurized heavy-water reactors with features of pressurized light-water-cooled reactors; from CANDU, it takes the heavy-water moderator, which gives the design an improved neutron economy that allows it to burn a variety of fuels; however, it replaces the heavy-water cooling loop with one containing conventional light water, greatly reducing costs; the name refers to its design power in the 1000 MWe class, with the baseline around 1200 MWe ↦ CANDU; heavy water; neutron economy **D** *ACR* **F** *ACR* **Pl** *ACR* **Sv** *ACR*

actinide *mat* • element with atomic number Z in a range from 89 (actinium) to 103 (lawrencium) ↦ minor actinide; major actinide; atomic number **D** *Aktinide* **F** *actinide* **Pl** *aktynowiec* **Sv** *aktinid; aktinoid*

actinium *mat* • radioactive chemical element denoted Ac, with atomic number $Z=89$, relative atomic mass $A_r=227.03$, density 10.07 g/cm^3, melting point 1051 °C, boiling point 3198 °C, and crustal average abundance 5.5×10^{-10} mg/kg; *a.* gave the name to the actinide series ↦ actinide **D** *Aktinium* **F** *actinium* **Pl** *aktyn* **Sv** *aktinium*

action level *rdp* • (for radiation protection situation where a radiation source is not under administrative control:) level of the absorbed dose or the concentration of radioactivity, which, in case it is exceeded, causes steps against other items than the reactivity source, e.g., after an unintentional release of radioactive material, a decision is made that prohibits usage of contaminated food ↦ absorbed dose; radioactive material; radiation source **D** *kritischer Schwellwert* **F** *niveau d'intervention* **Pl** *poziom działania* **Sv** *åtgärdsnivå*

activation *rdy* • process of creation of artificial radioactivity of a material through bombardment with neutrons ↦ artificial radioactivity; neutron **D** *Aktivierung* **F** *activation* **Pl** *aktywacja* **Sv** *aktivering*

activation cross section *xr* • cross section for creation of a radioactive nuclide due to certain nuclear reaction ↦ cross section; nuclide **D** *Aktivierungsquerschnitt* **F** *section efficace d'activation* **Pl** *przekrój czynny na aktywację* **Sv** *aktiveringstvärsnitt*

activation detector *mt* • radiation detector whose radioactivity arising from activation is used to determine the particle fluence or particle flux density ↦ detector; radiation; particle fluence; particle flux density **D** *Aktivierungsdetektor* **F** *détecteur par activation* **Pl** *detektor aktywacyjny* **Sv** *aktiveringsdetektor*

activation foil *mt* • activation detector which has a shape of a thin foil ↦ activation detector; radiation **D** *Aktivierungsfolie* **F** *feuille d'activation* **Pl** *folia aktywacyjna* **Sv** *aktiveringsfolium; aktiveringsfolie*

activation product *rdy* • substance that becomes radioactive due to irradiation, usually by neutrons ↦ radioactive; exposure[1]; neutron **D** *Aktivierungsprodukt* **F** *produit d'activation* **Pl** *produkt aktywacji* **Sv** *aktiveringsprodukt*

activity *rdy* • (in a given amount of material:) number of nuclei which decay per unit time; the SI unit of *a.* is the becquerel (Bq); an older unit of *a.*, which is still sometime encountered, is the curie (Ci) ↦ becquerel; curie; radioactivity **D** *Aktivität* **F** *activité* **Pl** *aktywność* **Sv** *aktivitet[1]*

activity concentration *rdy* • (for a given radioactive material:) ratio of the activity of a radioactive sample and the volume of the sample expressed in, e.g., Bq/m^3 ↦ specific activity; becquerel; radioactivity; radioactive material **D** *Aktivitätskonzentration* **F** *activité volumique* **Pl** *gęstość aktywności promieniowania* **Sv** *aktivitetstäthet; aktivitetskoncentration*

actual quality ↦ *steam quality*

actuation circuit *roc* • logic circuit that initiates an automatic action based on a signal from a nuclear power plant ↦ nuclear power plant **D** *Erregerkreis* **F** *circuit de commande* **Pl** *obwód uruchomiający* **Sv** *utlösningskrets; villkorskrets*

adjoint flux *rph* • (also called *adjoint of the neutron flux density*:) solution of the adjoint diffusion equation or adjoint transport equation; for a critical system, the *a.f.* is proportional to the importance function ↦ diffusion equation; transport equation; neutron flux; importance function **D** *Adjungierte der Neutronenflußdichte* **F** *adjoint de la densité de flux neutronique; adjoint du débit de fluence neutronique; flux adjoint* **Pl** *strumień sprzężony neutronów* **Sv** *adjungerad neutronflödestäthet*

adjoint of the neutron flux density ↦ *adjoint flux*

ADS ↦ *accelerator-driven system*

Advanced CANDU reactor ↦ *ACR*

advanced gas-cooled reactor *rty* • (abbreviated AGR:) nuclear reactor with graphite moderator in which gas is used as the coolant, the fuel is enriched and the fuel cladding is made of stainless steel ↦ gas-cooled reactor; carbon; coolant; moderator; enriched fuel[1] **D** *fortgeschrittener gasgekühlter Reaktor* **F** *réacteur avancé refroidi par gaz* **Pl** *za-*

awansowany reaktor chłodzony gazem **Sv** *avancerad gaskyld reaktor*

advantage factor *rph* • ratio of a certain radiation quantity value at a point where an elevated effect of some kind is achieved and the same quantity value at a reference point ↦ radiation   **D** *Überhöhungsfaktor* **F** *facteur d'avantage* **Pl** *współczynnik spiętrzenia* **Sv** *förhöjningsfaktor*

aerodynamic separation process ↦*enrichment*[1]

Ag ↦*silver*

age *rph* • (also called *Fermi age:*) parameter defined for slowing-down neutrons with energy E as $\tau \equiv \int_{E}^{E_0} \frac{D}{\xi \Sigma_s E'} dE'$, where D - diffusion coefficient, Σ_s - scattering cross section, ξ - average logarithmic energy decrement, E_0 - initial (source) energy; using a solution of the age equation for a point source of fast monoenergetic neutrons undergoing continuous slowing down in a nonabsorbing medium, it can be shown that $\tau = \frac{1}{6}\overline{r_s^2}$; here r_s is a mean square slowing-down distance, which is physically equivalent to the crow-flight distance from the point where a neutron is emitted as a fast neutron to the point where it slows down to energy E ↦ slowing-down   **D** *Fermi-Alter* **F** *âge* **Pl** *wiek* **Sv** *ålder; Fermi-ålder*

age of waste *wst* • **1.** (related to used fuel, activation products or waste from fuel reprocessing:) time after the irradiation was ended **2.** (related to waste which originates from contact with radioactive material:) time after separation from the radioactive material ↦ radioactive material   **D** *Abfallalter* **F** *âge des déchets* **Pl** *wiek odpadów* **Sv** *avfallsålder*

AGR ↦*advanced gas-cooled reactor*

air monitor *rdp* • radiation monitor for air-borne radioactive material ↦ radiation monitor   **D** *Luftmonitor; Luftüberwachungsgerät* **F** *moniteur atmosphèrique* **Pl** *wskaźnik radioaktywności powietrza* **Sv** *luftmonitor*

Al ↦*aluminum*

ALARA *rdp* • (acronym for *As Low as Reasonably Achievable:*) approach in safety radiation protection according to which all radiation due to usage of radiation sources shall be reduced as much as possible, taking into consideration the economic and social factors ↦ radiation source; radiation protection   **D** *ALARA* **F** *ALARA* **Pl** *ALARA* **Sv** *ALARA*

albedo *nap* • coefficient of reflection of radiation from a surface ↦ radiation   **D** *Albedo* **F** *albédo* **Pl** *albedo* **Sv** *albedo*

albedo dosimeter *rdp* • individual dosimeter which registers neutrons scattered from the body ↦ individual dosimeter   **D** *Albedodosimeter* **F** *dosimètre à albédo* **Pl** *albedometr* **Sv** *albedodosimeter*

alert ↦*reference level for emergency action*

alpha-bearing waste *wst* • wastes which contain one or several alpha emitters, usually actinides, in such amounts that the prescribed limits are exceeded ↦ actinide; alpha emitter   **D** *radioaktiver Abfall mit Alphastrahlern* **F** *déchet radioactif alpha* **Pl** *odpad alfa-promieniotwórczy* **Sv** *alfaavfall*

alpha box *rdp* • glove box specially designed for work with alpha emitters ↦ alpha emitter   **D** *Alphakasten* **F** *boîte à gants pour émetteurs alpha* **Pl** *komora rękawicowa do pracy z emiterem promieniowania alfa* **Sv** *alfabox*

alpha decay *rdy* • radioactive decay in which the alpha particle is emitted by an unstable nucleus ↦ alpha emitter; decay; radiation; nucleus   **D** *Alphazerfall* **F**

désintégration alpha **Pl** *rozpad alfa* **Sv** *alfasönderfall*

alpha emitter *rdy* • radioactive nuclide that is emitting alpha particles ↦ alpha particle; nuclide **D** *Alphastrahler* **F** *émetteur alpha* **Pl** *emiter promieniowania alfa* **Sv** *alfastrålare*

alpha particle *rd* • (denoted α:) particle with a positive electric charge identical with the nucleus of helium consisting of two protons and two neutrons, with rest mass 6.644 657 230(82) × 10^{-27} kg, typically produced in the process of alpha decay ↦ alpha decay; alpha radiation **D** *Alphateilchen* **F** *particule alpha* **Pl** *cząstka alfa* **Sv** *alfapartikel*

alpha radiation *rd* • (also called *alpha rays*:) alpha particles emitted during the alpha decay process; the range of the *a.r.* depends on the source of the radiation and in the air it varies from 25 mm for ^{232}Th to 86 mm for ^{212}Po ↦ radiation; beta radiation; gamma radiation; ionizing radiation **D** *Alphastrahlung* **F** *rayonnement alpha; rayons alpha* **Pl** *promieniowanie alfa* **Sv** *alfastrålning*

alpha ratio *rph* • (denoted α, also called *capture-to-fission ratio*:) ratio of the microscopic cross section for radiative capture σ_γ to the microscopic cross section for fission σ_f in a fissile material, $\alpha = \sigma_\gamma/\sigma_f$ ↦ cross section; radiative capture; fission **D** *Verhältniszahl α bei spaltbaren Kernen* **F** *facteur alpha* **Pl** *współczynnik α* **Sv** *infångningskvot*

alpha rays ↦*alpha radiation*

aluminum *mat* • chemical element denoted Al, with atomic number Z=13, relative atomic mass A_r=26.981538, density 2.70 g/cm^3, melting point 660.32 °C, boiling point 2519 °C; *a.* has reasonably high melting point and low thermal-neutron

cross section (0.23 b) to be suitable (to some extent) as a structure material in nuclear reactors (e.g. in magnox), but its low creep strength makes it unsuitable for temperatures above 300 °C↦ element; magnox; boral **D** *Aluminium* **F** *aluminium* **Pl** *glin; aluminium* **Sv** *aluminium*

ambient dose equivalent *rdp* • dose equivalent at 10 mm depth in a sphere with 30 cm in diameter made of tissue-like material with density of 1000 kg/m^3; the *a.d.e.* is measurable and is the most reliable approximation of the effective dose equivalent in an unknown radiative environment ↦ dose equivalent; effective dose equivalent **D** - **F** *équivalent de dose ambiant* **Pl** *środowiskowy równoważnik dawki pochłoniętej* **Sv** *miljödosekvivalent*

americium *mat* • synthetic chemical element denoted Am, with atomic number $Z = 95$, belonging to the family of minor actinides, created in nuclear reactors due to an irradiation of uranium and plutonium by neutrons ↦ minor actinide; transmutation; plutonium; uranium **D** *Americium* **F** *américium* **Pl** *ameryk* **Sv** *americium*

angle-integrated neutron flux ↦*neutron flux density*

angular cross section *xr* • differential cross section with respect to the spatial direction angle ↦ cross section; differential cross section **D** *raumwinkelbezogener Wirkungsquerschnitt* **F** *section efficace différentielle angulaire* **Pl** *przekrój czynny kątowy* **Sv** *vinkeltvärsnitt*

angular flux density ↦*neutron flux density*

annular flow *th* • two-phase flow pattern in which the liquid phase flows along channel walls as a liquid film, and also as droplets in the central part of the channel, which is mainly filled with the gaseous phase $\mapsto$ two-phase flow; two-phase flow pattern; bubbly flow; slug flow; mist flow **D** - **F** *écoulement annulaire* **Pl** *przepływ pierścieniowy* **Sv** *ringflöde*

anticipated transient without scram *rs* ● (abbreviated *ATWS:*) failure of the reactor protection (or shutdown) system when a transient required a reactor trip, leading potentially to core meltdown; $\mapsto$ reactor trip; transient; protective system **D** *ATWS-Störfall* **F** *transitoire sans arrêt d'urgence* **Pl** *stan przejściowy bez awaryjnego wyłączenia reaktora* **Sv** *transient med uteblivet reaktorsnabbstopp*

AP1000 reactor *rty* ● designed by Westinghouse, two-loop advanced pressurized water reactor configuration with capability of producing over 1000 MWe, employing extensive passive safety systems; each circulation loop consists of one steam generator and two circulation pumps; AP1000 is designed to use 17×17 pin array fuel assemblies containing mixed-oxide (PuO_2/UO_2) fuel; the first AP1000 reactor was constructed in China and connected to the grid on July 2, 2018 $\mapsto$ pressurized water reactor; passive safety **D** *AP1000-Reaktor* **F** *réacteur AP1000* **Pl** *reaktor AP1000* **Sv** *AP1000-reaktor*

approach to criticality *rph* ● series of small reactivity increases in a subcritical system by successive small changes of one of the system's parameters; through *a.t.c.* and using extrapolation, it is possible to predict the exact value of the parameter for which the system will be critical $\mapsto$ criticality **D** *Annäherung an den kritischen Zustand* **F** *approache sous-critique* **Pl** *dochodzenie do krytyczności* **Sv** *närmande till kriticitet*

Ar $\mapsto$ argon

$\mathbf{A}_r$ $\mapsto$*relative atomic mass*

area monitor *rdp* ● radiation monitor to survey the level of ionizing radiation at a given location, for example, in a nuclear reactor's surroundings $\mapsto$ ionizing radiation; radiation monitor **D** *Raummonitor; Raumüberwachungsgerät* **F** *moniteur de zone* **Pl** *monitor strefowy* **Sv** *omgivningsmonitor*

argon *mat* ● chemical element denoted Ar, with atomic number $Z=18$, relative atomic mass $A_r=39.948$, density 1.396 g/cm^3, melting point -189.35 °C, boiling point -185.85 °C, crustal average abundance 3.5 mg/kg and ocean abundance 0.45 mg/L; *a.* is the most common noble gas and its radioisotope ^{41}Ar, produced by the $^{40}Ar(n,\gamma)^{41}Ar$ reaction during the irradiation of air by neutrons, is the only significant radioisotope released to the atmosphere during normal reactor operation $\mapsto$ element; radioisotope **D** *Argon* **F** *argon* **Pl** *argon* **Sv** *argon*

artificial radioactivity *rdy* ● radioactivity of isotopes obtained artificially, first discovered by Irène and Frédéric Joliot-Curie in 1934 $\mapsto$ radioactivity; synthetic element; element **D** *künstliche Radioaktivität* **F** *radioactivité artificielle* **Pl** *promieniotwórczość sztuczna* **Sv** *konstgjord radioaktivitet*

asymptotic neutron flux density *rph* ● neutron flux density inside a volume filled with a certain substance at a distance equivalent to several mean free paths from the volume boundary, the local neutron sources or the local neutron absorbers $\mapsto$ neutron flux density; mean free path; absorber **D** *asymptotische Neutronenflussdichte* **F** *densité de flux asymptotique de neutrons* **Pl** *asymptotyczna gęstość strumienia neutronów* **Sv** *asymptotisk neutronflödestäthet*

atom *bph* ● smallest constituent unit

of a chemical element with mass in range from 1.67×10^{-27} to 4.52×10^{-25} kg and diameter from 62 pm (He) to 520 pm (Cs), containing electrons and a nucleus composed of neutrons and protons ↦ element; neutron; proton; electron **D** *Atom* **F** *atome* **Pl** *atom* **Sv** *atom*

atom density ↦*atomic number density*

atomic *bph* • in relation to an atom, e.g., atomic mass, atomic nucleus ↦ atomic mass; atomic nucleus; atomic number; atomic weight **D** *atom-* **F** *atomique* **Pl** *atomowa* **Sv** *atom-*

atomic mass *bph* • mass of a single neutral atom, often expressed in the non-SI atomic mass unit dalton (symbol Da or u); the rest mass of an atom X, with atomic number Z and mass number A is denoted $M({}_{Z}^{A}X)$ while that of its nucleus is denoted $m({}_{Z}^{A}X)$; the average of the atomic masses of all the isotopes of a chemical element, weighted by their respective abundance on Earth, is called the *atomic weight* ↦ atomic mass unit **D** *Atommasse* **F** *masse atomique* **Pl** *masa atomowa* **Sv** *atommassa*

atomic mass number ↦*mass number*

atomic mass unit *bph* • (also called *dalton*, denoted Da or u:) mass unit used in physics and chemistry that corresponds exactly to one-twelfth of the mass of an unbound neutral atom of ^{12}C in its nuclear and electronic ground state and at rest, that is, 1 u = 1 Da = 1.660 539 066 60(50)$\times 10^{-27}$ kg = 931.494 102 42(28) MeV/c^2 ↦ atomic mass **D** *Atommasseneinheit* **F** *unité de masse atomique* **Pl** *jednostka masy atomowej* **Sv** *atommassenhet; universell massenhet*

atomic nucleus *bph* • central region of an atom consisting of neutrons and protons, with a diameter from 1.7566×10^{-27} m for hydrogen to 11.7142×10^{-27} m for uranium; since a stable nucleus has approximately a constant density, its radius R can be approximately computed as $R = r_0 A^{1/3}$, where A - atomic mass number and $r_0 = 1.25 \times 10^{-15}$ m ↦ neutron; proton; mass number; mass defect **D** *Atomkern* **F** *noyau atomique* **Pl** *jądro atomu* **Sv** *atomkärna*

atomic number *bph* • (denoted Z:) characteristic number describing a chemical element, equal to the number of protons found in the nucleus of every atom of that element, identical to the charge number of the atomic nucleus ↦ mass number; nucleon; element; atomic nucleus **D** *Atomnummer; Ordnungszahl* **F** *numéro atomique* **Pl** *liczba atomowa* **Sv** *atomnummer*

atomic number density *bph* • number of atoms of a given type per unit volume; for a substance with mass density ρ (kg/m^3) and molar mass M (kg/mol), the *a.n.d.* is found as $N = 10^3 \rho N_A / M$ atoms/m^3, where N_A is Avogadro's constant ↦ Avogadro's constant **D** - **F** - **Pl** *gęstość liczbowa atomów* **Sv** -

atomic weight ↦*atomic mass*

attenuation *rd* • reduction of a radiation quantity (such as, e.g., radiation flux density or radiation energy density) when passing through matter due to all kinds of interactions with the matter (thus reduction due only to increasing distance from the radiation source is not included) ↦ radiation; radiation source **D** *materielle Schwächung* **F** *atténuation* **Pl** *osłabienie* **Sv** *dämpning*

ATWS ↦*anticipated transient without scram*

augmentation distance ↦*linear extrapolation distance*

autocatalytic instability *rs* • attribute of a nuclear reactor in which a power increase can sometime cause increase of the reactivity, which, in turn, causes an additional increase of the reactor power, leading to reactor excursion, which is stopped when the reactor core is destroyed ↦ reactivity; reactor excursion; reactivity coefficient; reactivity feed-back **D** *selbstverstärkende Instabilität* **F** *instabilité autocatalytique* **Pl** *samowzmacniająca się niestabilność* **Sv** *självförstärkande instabilitet*

automatic depressurization *rs* • relief of vapour through a pressure relief system in a boiling water reactor, to allow pumping of water to the core by the emergency core cooling system at low reactor pressure ↦ boiling water reactor; emergency core cooling system; pressure relief system **D** *automatiche Druckentlastung* **F** *dépressurisation automatique* **Pl** *wymuszony upust pary; automatyczny upust pary* **Sv** *tvångsnedblåsning*

availability factor *roc* • (for a reactor unit:) ratio of the time during which a generator in a reactor unit is connected to the electrical grid to the total calendar time during a given period of time, specified in percent ↦ capacity factor **D** *Verfügbarkeitsfaktor* **F** *taux de disponibilité; coefficient de disponibilité* **Pl** *wskaźnik dyspozycyjności* **Sv** *tillgänglighetsfaktor; tidtillgänglighetsfaktor*

average logarithmic energy decrement *rph* • (denoted ξ:) average of a natural logarithm of a ratio of neutron energy, subject to reduction per single collision from E_1 to E_2, given as:

$$\xi \equiv \overline{\ln \frac{E_1}{E_2}} = 1 + \frac{\alpha}{1 - \alpha} \ln \alpha,$$

where $\alpha = [(A - 1)/(A + 1)]^2$ and A is the mass number of the species ↦ energy decrement; scattering; mass number

D *mittleres logarithmisches Energidekrement* **F** *décrément logarithmique moyen de l'energie* **Pl** *średni logarytmiczny dekrement tłumienia* **Sv** *logaritmiskt energidekrement*

average power *roc* • ratio of a produced energy and the time during which the energy was produced ↦ availability factor; capacity factor **D** *mittlere Leistung* **F** *puissance moyenne* **Pl** *moc średnia* **Sv** *medeleffekt*

average scattering angle cosine *rph* • quantity appearing in the derivation of the one-speed diffusion equation and defined as $\overline{\mu}_0 \equiv \langle \mathbf{\Omega} \cdot \mathbf{\Omega}' \rangle$, where $\mathbf{\Omega}$ and $\mathbf{\Omega}'$ are unit direction vectors of a neutron before and after scattering, respectively, and $\langle \rangle$ represents averaging over all scattering directions; for elastic scattering from stationary nuclei with mass number A, the *a.s.a.c.* is found as $\overline{\mu}_0 = 2/(3A)$ ↦ diffusion equation; scattering **D** - **F** - **Pl** *średni kosinus kąta rozpraszania* **Sv** *medelriktningscosinen*

Avogadro's constant *bph* • (usually denoted N_A:) one of the exact fundamental physical constants equal to $N_A = 6.022\ 140\ 76 \times 10^{23}$ mol^{-1}, used to define the SI base unit of the amount of substance, mole ↦ mole; Avogadro's number **D** *Avogadro-Konstante* **F** *constante d'Avogadro* **Pl** *stała Avogadra* **Sv** *Avogadros konstant*

Avogadro's number *bph* • (usually denoted N_0:) dimensionless number equal to the number of elementary entities (atoms or molecules) in one mole, with a fixed numerical value of Avogadro's constant ↦ mole; Avogadro's constant **D** *Avogadro-Zahl* **F** *nombre d'Avogadro* **Pl** *liczba Avogadra* **Sv** *Avogadros tal*

axial fuel gap *nf* • (also called *fuel gap*:) axial distance between neighbouring fuel pellets ↦ fuel pellet **D**

axialer Brennstabzwischenraum **F** *intervalle des barreaux de combustible axials; écart axial entre les barreaux de combustible* **Pl** *osiowy odstęp między pastylkami paliwowymi* **Sv** *kutsavstånd*

axial offset anomaly ↦*crud*

axial peaking factor *th* ● (in a reactor with vertical fuel assemblies, also called *axial shape factor*:) ratio of the maximum local power density to the averaged power density in the vertical direction in the whole core (this defines the core *a.p.f.*) or in a single assembly (this defines the assembly *a.p.f.*) ↦ radial peaking factor **D** *axialer Formfaktor* **F** *facteur (de forme) axial* **Pl** *osiowy współczynnik rozkładu mocy* **Sv** *axiell formfaktor*

axial shape factor ↦*axial peaking factor*

Bb

B ↦boron

BA ↦burnable absorber

back-end of the fuel cycle *nf* • last stage of the nuclear fuel cycle including the interim storage of the used fuel, possible fuel re-processing and fuel waste management ↦ fuel cycle; front-end of the fuel cycle **D** *Ende des Brennstoffkreislaufs* **F** *fin du cycle du combustible* **Pl** *etap końcowy cyklu paliwowego* **Sv** *slutsteg²*

backfitting ↦retrofitting

background *mt* • (in radiation measurement:) signals of other origin than the radiation that is to be detected ↦ detector **D** *Hintergrund; Untergrund* **F** *bruit de fond; mouvement propre* **Pl** *tło* **Sv** *backgrund*

background radiation *rd* • radiation other than the one that is to be measured, irrespective of the source of radiation; there are several natural sources of the *b.r.* such as the cosmic radiation or the radiation caused by radioactive isotopes naturally occurring on Earth ↦ radiation; radioisotope **D** *Hintergrundstrahlung* **F** *fond de rayonnement* **Pl** *promieniowanie tła* **Sv** *backgrundsstrålning*

backpressure power *th* • electric power generated in a turbogenerator in which the exit steam is used for other purposes, typically for heating ↦ turbogenerator **D** *Gegendruckleistung; Gegendruckskraft* **F** *puissance de contre-pression* **Pl** *moc przeciwprężna* **Sv** *mottryckseffekt*

back-scattering *nap* • scattering of radiation with an angle greater than 90° from the initial radiation direction ↦ scattering **D** *Rückstreuung* **F** *rétrodiffusion* **Pl** *rozpraszanie wsteczne* **Sv** *bakåtspridning*

backwash (GB) ↦stripping (US)

bare reactor *rty* • nuclear reactor in which there is no reflector surrounding the core; the *b.r.* assumption is frequently adopted in comparative theoretical studies of bucklings and neutron flux distributions in reactors of various shapes ↦ buckling; reflector; reflector saving **D** *reflektorloser Reaktor; nackter Reaktor* **F** *réacteur nu; réacteur sans réflecteur* **Pl** *reaktor bez reflektora* **Sv** *bar reaktor*

barn *xr* • (denoted b:) unit used in the nuclear physics for measuring the microscopic cross section of a given substance for a specific type of a nuclear reaction: 1 barn = 1 b = 10^{-28} m² = 10^{-24} cm² ↦ microscopic cross section; nuclear reaction **D** *Barn* **F** *barn* **Pl** *barn* **Sv** *barn*

barrier *wst* • natural or constructed device, which delays or prevents dispersion of radioactive nuclides; a constructed barrier is a device that has been changed or manufactured by humans ↦ nuclide; radioactive **D** *Barriere* **F**

barrière de confinement; barrière **Pl** bariera **Sv** barriär

barring ↦*turning*

barytes concrete *mat* • high-density concrete, containing barytes (white or colourless minerals consisting of barium sulphate $BaSO_4$) for better shielding properties ↦ shielding; attenuation **D** *Barytbeton* **F** *béton baryté* **Pl** *beton barytowy; barytobeton* **Sv** *barytbetong*

base load ↦*base power*

base-load operation *roc* • operation at a constant, high power ↦ load following **D** *Grundlastbetrieb* **F** *fonctionnement en base* **Pl** *praca w obciążeniu podstawowym* **Sv** *baslastdrift*

base-load station *roc* • power plant designed for a base-load operation ↦ base-load operation; base power **D** *Grundlastkraftwerk* **F** *centrale fonctionnant en base; centrale (de puissance) de base* **Pl** *elektrownia pracująca w obciążeniu podstawowym* **Sv** *baslaststation*

base power *roc* • (also called *base load:*) minimum level of demand on an electrical grid over a longer period of time, usually met by high-efficiency power plants ↦ base-load operation; peak power **D** *Grundleistung* **F** *puissance de creux* **Pl** *moc podstawowa* **Sv** *botteneffekt; baseffekt*

batch *nf* • amount of nuclear material that, due to bookkeeping reasons, is treated as a unit at the measurement point in a nuclear installation ↦ key measurement point; safeguards **D** *Charge* **F** *lot* **Pl** *wsad* **Sv** *sats*

batch-by-batch method *nf* • method to calculate the fuel-cycle costs in a nuclear reactor in which costs are added and incomes subtracted for each fuel load that is placed in the reactor ↦ fuel cycle **D** *Chargenmethode* **F** *méthode par lots* **Pl**

obliczanie wsadowe (kosztów) **Sv** satsvis beräkning

Be ↦*beryllium*

beam *rd* • current of particles directed around a narrow angle in space ↦ particle; beam hole; beam trap **D** *Strahl* **F** *faisceau* **Pl** *wiązka (promieniowania)* **Sv** *stråle*

beam hole *rd* • channel that, for experimental purposes, releases radiation through a radiation shield ↦ radiation shield **D** *Strahlrohr* **F** *canal expérimental à sortie de faisceau* **Pl** *kanał doświadczalny* **Sv** *strålkanal*

beam trap *rd* • device designed to stop a beam of particles and to reduce the secondary radiation to an acceptable level due to, e.g., scattering ↦ beam **D** *Strahlenfänger* **F** *piège à faisceau* **Pl** *pułapka wiązki promieniowania* **Sv** *strålfång*

becquerel *rd* • SI unit used for activity, defined as one transformation per second; the *b.* replaced an older unit of activity, and one which is still in use, the curie (Ci) ↦ curie; radioactivity; decay **D** *Becquerel* **F** *becquerel* **Pl** *bekerel* **Sv** *becquerel*

bedrock depository *wst* • underground installation placed in a bedrock and used for the ultimate disposal of radioactive wastes; according to the *KBS-3 technology*, the waste disposal process involves placement of wastes into a boron steel canister and enclosure in a copper capsule, which is subsequently placed in a hole in the *b.d.* and overpacked with bentonite clay ↦ ultimate waste disposal; geological repository **D** *Endlager im Granit* **F** *dépôt*[1] **Pl** *składowisko ostateczne w podłożu skalnym* **Sv** *bergförvar*

beginning of cycle *roc* • period during reactor operation just after the refuelling, used as a reference point in the analysis of the fuel in the reactor ↦ refuelling; end of cycle; operating cycle; operating period **D** *Zyklusanfang* **F** *début de cycle* **Pl** *początek kampanii paliwowej* **Sv** *början av driftperiod*

beryllium *mat* • chemical element denoted Be, with atomic number $Z=4$, relative atomic mass $A_r=9.012182$, density 1.85 g/cm^3, melting point 1287 °C, boiling point 2471 °C, crustal average abundance 2.8 mg/kg and ocean abundance 5.6×10^{-6} mg/L; the *b.* metal is an excellent material for use as a moderator or reflector (from the neutronic standpoint, superior to graphite, with the macroscopic cross section for absorption $\sigma_a = 0.0092$ b and for scattering $\sigma_s = 6.14$ b), but it is relatively brittle and expensive, and at present there seems little prospect that it will be used to any extent ↦ element; graphite; moderator **D** *Beryllium* **F** *béryllium* **Pl** *beryl* **Sv** *beryllium*

Bessel functions of the first kind *mth* • Bessel functions $J_n(x)$, where n is the order of the function; $J_0(x)$ describes the radial neutron flux distribution in a cylindrical core as obtained from a solution of the neutron diffusion equation; in general $J_n(x)$ can be represented as the following series,

$$J_n(x) = \left(\frac{x}{2}\right)^n \sum_{k=0}^{\infty} \frac{(-1)^k \left(\frac{x}{2}\right)^{2k}}{k!(n+k)!};$$

for small x the following approximations are valid $J_n(x) \approx \frac{1}{2^n n!} x^n$, $J_0(x) \approx 1 - \frac{x^2}{4} + \frac{x^4}{64} - \frac{x^6}{2304}$, and $J_1(x) \approx \frac{x}{2} - \frac{x^3}{16} + \frac{x^5}{384}$; first three roots of $J_0(x) = 0$ are: $x_1 \approx 2.40483$, $x_2 \approx 5.52008$, $x_3 \approx 8.65373$ ↦ Bessel functions of the second kind; Bessel functions of the third kind **D** *Bessel-Funktionen*

erster Gattung **F** *fonctions de Bessel de première espèce* **Pl** *funkcje Bessela pierwszego rodzaju* **Sv** *besselfunktionerna av första slaget*

Bessel functions of the second kind *mth* • Bessel functions $Y_n(x)$, where n is the order of the function; $Y_0(x)$ describes the radial neutron flux distribution in a hollow cylindrical region as obtained from a solution of the neutron diffusion equation; for small x, the following approximations are valid, $Y_0(x) \approx \frac{2}{\pi} \left[\left(\gamma + \ln \frac{x}{2}\right) J_0(x) + \frac{x^2}{4} \right]$ and $Y_1(x) \approx \frac{2}{\pi} \left[\left(\gamma + \ln \frac{x}{2}\right) J_1(x) - \frac{1}{x} \right]$ where $J_0(x)$ and $J_1(x)$ are Bessel functions of the first kind and zero and first order, respectively, and $\gamma = 0.577216$ ↦ Bessel functions of the first kind; Bessel functions of the third kind **D** *Bessel-Funktionen zweiter Gattung* **F** *fonctions de Bessel de deuxième espèce* **Pl** *funkcje Bessela drugiego rodzaju* **Sv** *besselfunktionerna av andra slaget*

Bessel functions of the third kind *mth* • (also called *Hankel functions:*) functions $H_n^{(1)}(x) = J_n(x) + iY_n(x)$ and $H_n^{(2)}(x) = J_n(x) - iY_n(x)$, where n is the order of the function, $J_n(x)$ is the Bessel function of the first kind and n-th order, $Y_n(x)$ is the Bessel function of the second kind and n-th order and $i = \sqrt{-1}$ ↦ Bessel functions of the first kind; Bessel functions of the second kind **D** *Bessel-Funktionen dreiter Gattung* **F** *fonctions de Bessel de troisième espèce* **Pl** *funkcje Bessela trzeciego rodzaju* **Sv** *besselfunktionerna av tredje slaget*

Bessel's differential equation *mth* • ordinary differential equation:

$$\frac{d^2y}{dx^2} + \frac{1}{x}\frac{dy}{dx} + \left(\alpha^2 - \frac{n^2}{x^2}\right) y = 0$$

with a general solution as follows

$y(x) = C_1 J_n(\alpha x) + C_2 Y_n(\alpha x)$, where C_1 and C_2 are constants, J_n is the Bessel function of the first kind and n-th order and Y_n is the Bessel function of the second kind and n-th order ↦ Bessel functions of the first kind; Bessel functions of the second kind; Bessel functions of the third kind **D** *Besselsche Differentialgleichung* **F** *équation différentielle de Bessel* **Pl** *równanie różniczkowe Bessela* **Sv** *Bessel differentialekvationen*

beta decay *rdy* • radioactive decay through emission of a beta particle, after which the atomic number of a decaying nucleus increases or decreases by one and the mass number is constant; in the former case a neutron in the nucleus changes to a proton, and an electron (β^-) together with anti-neutrino $(\bar{\nu})$ are emitted; in the latter case a proton in the nucleus changes to a neutron, and a positron (β^+) together with a neutrino (ν) are emitted ↦ alpha decay; beta particle; beta emitter; electron **D** *Betazerfall* **F** *désintégration bêta* **Pl** *rozpad beta* **Sv** *betasönderfall*

beta emission detector ↦*collectron*

beta emitter *rdy* • radioactive nuclide that is emitting beta particles ↦ beta decay; beta particle; nuclide **D** *Betastrahler* **F** *émetteur bêta* **Pl** *emiter beta* **Sv** *betastrålare*

beta particle *rd* • electron (β^-) or positron (β^+) emitted from a nuclide due to radioactive decay ↦ beta decay; beta emitter; nuclide **D** *Betateilchen* **F** *particule bêta* **Pl** *cząstka beta* **Sv** *betapartikel*

beta quenching *mat* • sudden cooling of zirconium alloy from high temperature, in which the β phase is stable ↦ zirconium **D** *Betaaushärtung* **F** *trempe bêta* **Pl** *hartowanie beta* **Sv** *betasläckning*

beta radiation *rd* • beta particles emitted during a radioactive beta decay ↦ beta decay; beta emitter; beta particle; beta emission detector **D** *Betastrahlung* **F** *rayonnement bêta; rayons bêta* **Pl** *promieniowanie beta* **Sv** *betastrålning*

beta treatment *mat* • treatment of uranium by heating, followed by sudden cooling, which leads to formation of uranium β-phase; the *b.t.* reduces risks of swelling, especially during irradiation ↦ irradiation; swelling **D** *Betabehandlung* **F** *traitement bêta* **Pl** *obróbka beta* **Sv** *betabehandling*

binding energy *bph* • (of nucleus:) energy equivalent to the work that has to be done to decompose a nucleus into the constituent nucleons; for a nucleus composed of Z protons and $N = A - Z$ neutrons, the *b.e.* can be calculated as $\mathrm{BE}(^A_Z X) = [ZM(^1_1 H) + (A - Z)m_n - M(^A_Z X)]c^2$, where m_n - rest mass of neutron, $M(^1_1 H)$ - rest mass of hydrogen atom, $M(^A_Z X)$ - rest mass of atom $^A_Z X$ and c - speed of light ↦ nucleus; nucleon; atomic mass **D** *Bindungsenergie* **F** *énergie de liaison* **Pl** *energia wiązania* **Sv** *bindningsenergi*

biological half-life *rdp* • time during which an amount of a certain substance in a biological system reduces to one-half due to biological processes; the *b.h.-l.* is used when the reduction of the substance approximately follows the exponential curve ↦ effective half-life; radioactive half-life **D** *Biologische Halbwertzeit* **F** *période biologique* **Pl** *biologiczny okres półrozpadu* **Sv** *biologisk halveringstid*

biological shield *rcs* • radiation shield whose main purpose is to reduce the radiation to the level that is not dangerous to humans ↦ thermal shield **D** *biologische Abschirmung; biologischer Schild* **F** *bouclier biologique* **Pl** *osłona* *biologiczna* **Sv** *biologisk strålskärm; biologisk skärm*

Biot's number *th* • dimensionless number representing a ratio of the heat transfer coefficient h to the thermal conductivity of the wall material λ, defined as $Bi = hL/\lambda$, where L is a characteristic length of the heat transfer system ↦ thermal conductivity; heat transfer coefficient   **D** *Biot-Zahl* **F** *nombre de Biot* **Pl** *liczba Biota* **Sv** *Biots tal*

bird cage *rs* • device to keep a minimum necessary distance between objects containing fissile material, to prevent criticality, when the objects come too close to each other ↦ criticality; fissile material   **D** *Abstandsgestell* **F** *cage de transport* **Pl** *klatka transportowa* **Sv** *säkerhetsställ*

bitumen *mat; wst* • asphalt used to infuse radioactive wastes ↦ radioactive waste   **D** *Bitumen* **F** *bitume* **Pl** *bitum* **Sv** *bitumen*

bituminization *wst* • binding of radioactive wastes through infusion in the bitumen ↦ bitumen; radioactive waste   **D** *Bituminierung* **F** *bitumage; enrobage par le bitumen* **Pl** *bitumizacja* **Sv** *bitumeningjutning; bituminering*

black *nap* • (concerning material or device in nuclear engineering:) which absorbs all neutrons of a certain energy, incident to this material or device ↦ absorption; gray   **D** *schwarz* **F** *noir* **Pl** *całkowicie pochłaniający; czarny* **Sv** *totalabsorberande*

blanket *rcs* • region in a nuclear reactor located in direct proximity to the core that contains a fertile material, which is converted into new nuclear fuel ↦ conversion; fertile material; reactor core   **D** *Brutzone* **F** *couche fertile; couverture* **Pl** *płaszcz (reaktora)* **Sv** *mantel*

blowdown *th* • first stage following a postulated break of a pipe in the primary system of a light-water nuclear reactor; the *b.* is usually divided into two parts: the blowdown of a subcooled water and the blowdown of saturated water ↦ loss-of-coolant accident   **D** *Ausblasen* **F** *évacuation* **Pl** *wydmuch* **Sv** *nedblåsning*

blowdown system *rcs* • part of the pressure relief system that is used during blowdown of steam to the wetwell ↦ blowdown; pressure relief system; wetwell   **D** *Ausblassystem* **F** *système de soufflage*[1] **Pl** *układ wydmuchu* **Sv** *nedblåsningssystem*

BOC ↦ *beginning of cycle*

body burden *rdp* • total activity of a certain radionuclide in a human body or in an animal ↦ activity; radionuclide   **D** *Körperbelastung* **F** *charge corporelle* **Pl** *aktywność ciała* **Sv** *kroppsaktivitet*

boiling crisis ↦ *critical heat flux*

boiling reactor *rty* • type of a nuclear reactor in which the primary coolant is boiling at the operational pressure ↦ light-water reactor; pressurized water reactor; saturated boiling   **D** *Siederaktor* **F** *réacteur bouillant* **Pl** *reaktor wrzący* **Sv** *kokarreaktor*

boiling water cooled reactor *rty* • boiling reactor with water as the primary coolant ↦ boiling reactor; light-water reactor; pressurized water reactor; saturated boiling   **D** *siedewassergekühlter Reaktor* **F** *réacteur refroidi par eau bouillante* **Pl** *reaktor chłodzony wrzącą wodą* **Sv** *kokvattenkyld reaktor*

boiling water reactor *rty* • (abbreviated *BWR:*) boiling reactor with water as the primary coolant and moderator ↦ boiling reactor; light-water reactor; pressurized water reactor; saturated boiling   **D** *Siedewasserreaktor* **F** *réacteur à eau bouillante* **Pl** *reaktor wodny wrzący; reaktor z wrzącą wodą* **Sv** *kokvattenreaktor; lättvattenkokare; BWR*

Boltzmann constant *bph* • (usually

denoted k_B:) one of the exact fundamental physical constants equal to $k_B = 1.380\ 649 \times 10^{-23}$ J/K, used in a definition of the kelvin; the *B.c.* relates the average relative kinetic energy of particles in a gas with the temperature of the gas ↦ kelvin **D** *Boltzmann-Konstante* **F** *constante de Boltzmann* **Pl** *stała Boltzmanna* **Sv** *Boltzmanns konstant*

bond[1] *mat* • close contact between nuclear fuel and cladding achieved either by mechanical or metallurgical way ↦ cladding **D** *Verbund* **F** *liaison[1]* **Pl** *wiązanie* **Sv** *bindning*

bond[2] *mat* • substance that is causing the bond ↦ bond[1] **D** *Verbundmaterial* **F** *liaison[2]* **Pl** *substancja wiążąca* **Sv** *bindämne*

bone-seeker *rdp* • substance that is absorbed in bones to a larger extent than in other living tissues ↦ dose **D** *Knochensucher* **F** *substance ostétrope* **Pl** *substancja osteotropowa* **Sv** *bensökande ämne*

book inventory *sfg* • balance of the latest physically determined inventory of the nuclear material and all booked changes of the material afterwards ↦ nuclear material **D** *Buchbestand* **F** *stock comptable* **Pl** *zaksięgowana ilość materiału rozszczepialnego* **Sv** *bokförd kärnämnesmängd; bokförd mängd*

booster element *nf* • fuel assembly which is introduced to a reactor core to compensate a temporary reactivity defect due to, e.g., the xenon transient ↦ fuel assembly; xenon transient **D** *Anfahrbrennelement* **F** *élément de dopage* **Pl** *kaseta wspomagająca* **Sv** *hjälppatron*

boral *mat* • solid dispersion of boron carbide in aluminium, used as a neutron absorber ↦ neutron absorber; aluminum; boron **D** *Boral* **F** *boral* **Pl** *boral* **Sv** *boral*

boron *mat* • chemical element de-noted B, with atomic number $Z=5$, relative atomic mass $A_r=10.811$, density 2.37 g/cm^3, melting point 2075 °C, boiling point 4000 °C, crustal average abundance 10 mg/kg and ocean abundance 4.44 mg/L; *b.* is a useful control material because the absorption cross section for neutrons is large over a considerable range of neutron energies, with the microscopic cross section for absorption of thermal neutrons $\sigma_a = 759$ b; boron carbide B$_4$C is preferred in boiling water reactors ↦ element; boral; boron control **D** *Bor* **F** *bore* **Pl** *bor* **Sv** *bor*

boron chamber *mt* • ionization chamber that contains boron or boron compounds, primarily dedicated to detection of slow neutrons ↦ ionization chamber **D** *Borkammer* **F** *chambre à bore* **Pl** *komora (jonizacyjna) borowa* **Sv** *borkammare*

boron control *roc* • type of reactor power control in a pressurized water-cooled reactor through changes of the concentration of boric acid (H$_3$BO$_3$) in the moderator ↦ power control **D** *Borsteuerung* **F** *contrôle par le bore* **Pl** *sterowanie borem* **Sv** *borstyrning*

boron counter tube *mt* • counter tube containing boron primarily dedicated to detection of slow neutrons ↦ boron chamber; counter tube **D** *Borzählrohr* **F** *tube compteur à bore* **Pl** *licznik borowy* **Sv** *borräknerör*

boron curtain ↦boron plate

boron equivalent *mat* • measure of contamination level of reactor material, in particular of nuclear fuel, indicating the neutron absorption ability due to the contamination, corresponding to an equivalent boron content ↦ radioactive contamination; neutron absorption; nuclear fuel; boron **D** *Boräquiv-*

alent **F** *équivalent bore* **Pl** *równoważnik borowy* **Sv** *boreqvivalent*

boron glass rod *rcs* • neutron absorber, formed as a glass rod containing boron, which can function as a burnable absorber ↦ boron plate; burnable absorber; neutron absorber **D** *Borglasstab* **F** *barre de bore vitrifiée* **Pl** *pręt ze szkła borowego* **Sv** *borglasstav*

boron injection *roc* • addition of a diluted boron to coolant in water moderated reactor to reduce the reactivity at the shutdown ↦ watermoderated reactor; reactivity; shutdown **D** *Boreinspritzung* **F** *injection de bore* **Pl** *wtrysk boru* **Sv** *borinsprutning*

boron plate *rcs* • neutron absorber, formed as a plate containing boron, which can function as a burnable absorber ↦ boron glass rod; burnable absorber; neutron absorber **D** *Borschild* **F** *plaque de bore* **Pl** *płyta borowa* **Sv** *borplåt*

borosilicate glass *mat* • glass that is used as a binding material for radioactive wastes ↦ radioactive waste **D** *Borsilikatglas* **F** *verre borosilicaté* **Pl** *szkło borokrzemianowe* **Sv** *borsilikatglas*

bottom nozzle *nf* • plate in the lower region of the PWR fuel assembly supporting fuel rods; for the BWR fuel assembly this part is called the *bottom tie plate* ↦ PWR; BWR; reactor core; top nozzle **D** *untere Gitterplatte* **F** *plaque-support inférieure* **Pl** *płyta dolna* **Sv** *bottenplatta*

bottom tie plate ↦ *bottom nozzle*

boundary layer *th* • region inside a fluid moving in a vicinity of a solid wall in which the viscosity forces are at least of the same order of magnitude as the inertial forces ↦ turbulent flow **D** *Fluiddynamische Grenzschicht* **F** *couche limite* **Pl** *warstwa przyścienna* **Sv** *gränsskikt*

bowing *nf* • bending of a fuel rod due

to, e.g., a non-uniform temperature distribution ↦ fuel rod **D** *Durchbiegen* **F** *arcure; déformation en arc* **Pl** *wyboczenie (pręta paliwowego)* **Sv** *stavböjning*

BR ↦ *breeding ratio*

Brayton cycle *th* • reference thermodynamic cycle taking place in gas turbines; the *B.c.* consists of the following processes: adiabatic compression, isobaric heating, adiabatic de-compression and isobaric cooling; next to the Rankine cycle, *B.c.* is the most widely implemented thermodynamic cycle in the electric power generation; assuming ideal processes and constant heat capacity of the gas, the *B.c.* efficiency is given as $\eta = 1 - (p_1/p_2)^{(\kappa-1)/\kappa}$, where $\kappa = c_p/c_v$ - ratio of specific heats at constant pressure and volume of the gas, p_1 - low pressure of the cycle, p_2 - high pressure of the cycle, $r = p_2/p_1$ - compression ratio ↦ Carnot cycle; Rankine cycle; cycle efficiency **D** *Joule-Kreisprozess* **F** *cycle de Brayton* **Pl** *obieg Braytona-Joule'a* **Sv** *Braytoncykel*

breeder reactor *rty* • nuclear reactor in which the number of new fissile atoms created by conversion is greater than the total number of fuel atoms used in the same period of time; this means that the conversion ratio in the *b.r.* is greater than 1, and thus, it is called the breeding ratio ↦ conversion; conversion ratio; breeding ratio **D** *Brutreaktor; Brütter* **F** *réacteur surrégénérateur; surrégénérateur* **Pl** *reaktor powielający* **Sv** *bridreaktor*

breeding *rph* • process in which the number of new fissile atoms created by conversion of a fertile atoms is greater than the total number of fuel atoms used in the same period of time; fertile isotopes ^{238}U and ^{232}Th can be converted into fissile isotopes ^{239}Pu

and ^{233}U, respectively, and can be used for the *b.* ↦ conversion; conversion ratio; breeding ratio **D** *Brüten* **F** *surrégénération* **Pl** *powielanie* **Sv** *bridning*

breeding gain *rph* ● (denoted G:) measure of a new fuel production in a breeder reactor defined as $G = BR - 1 = (F_E - F_B)/F_D = F_G/F_D$, where BR - breeding ratio, F_E - fissile inventory in the reactor at the end of a cycle, F_B - fissile inventory in the reactor at the beginning of a cycle, F_G - fissile material gained per cycle and F_D - fissile material destroyed per cycle ↦ breeder reactor; breeding ratio **D** *Brutgewinn* **F** *gain de surrégénération* **Pl** *uzysk powielania* **Sv** *bridvinst*

breeding ratio *rph* ● (abbreviated BR:) ratio of the number of fissile nuclei formed to the number destroyed in a reactor when it is larger than unity and breeding occurs; the *b.r.* determines the reactor ability to produce new fuel, expressed as $BR = \eta - (1 + L)$, where BR is the *b.r.*, η is the neutron yield per absorption of one neutron in fuel, L is the number of neutrons that are lost due to a parasitic absorption or escape from the reactor; since $L > 0$ and $BR > 1$, thus it must be $\eta > 2$, which constitutes the fundamental criterion for the fuel in the breeding reactor to make the breeding possible ↦ conversion; conversion ratio; neutron yield per absorption; breeding gain **D** *Brutverhältnis* **F** *rapport de surrégénération* **Pl** *współczynnik powielania* **Sv** *bridförhållande*

Breit-Wigner formula *xr* ● formula describing the microscopic cross section for radiative capture σ_γ in the neighbourhood of a single isolated resonance around energy E_0 as follows: $\sigma_\gamma(E_c) = \sigma_0 (\Gamma_\gamma/\Gamma) (E_0/E_c)^{1/2} / (1 + y^2)$,

where $y = 2(E_c - E_0)/\Gamma$; here σ_0 - total cross section at the resonance energy E_0, E_c - center of mass energy, Γ - so-called total line width (characterizing the full width at half-maximum of the resonance), Γ_γ - radiative line width (characterizing the probability that the compound nucleus will decay via gamma emission) ↦ microscopic cross section; resonance **D** *Breit-Wigner-Formel* **F** - **Pl** *wzór Breita-Wignera* **Sv** *Breit-Wigner formeln*

bremsstrahlung *rd* ● photons emitted by charged particles when they are deflected by the electric fields of nuclei and ambient electrons ↦ radiation; gamma radiation **D** *Bremsstrahlung* **F** *rayonnement de freinage* **Pl** *promieniowanie hamowania* **Sv** *bromsstrålning*

bubbly flow *th* ● two-phase flow pattern, in which the gas phase exists as small bubbles moving in a continuous liquid phase ↦ two-phase flow; two-phase flow pattern **D** *blasenströmung* **F** *écoulement à bulles* **Pl** *przepływ pęcherzykowy* **Sv** *bubbelströmning*

buckling *rph* ● (denoted B^2:) approximate measure of a square of a curvature of the surface which determines how the neutron flux density varies with position (e.g., in a nuclear reactor); for a bare critical reactor the geometric *b.* and the material *b.* are equal ↦ neutron flux density; bare reactor; critical; critical equation **D** *Flußwölbung* **F** *laplacien* **Pl** *parametr krzywizny (strumienia neutronów w reaktorze)* **Sv** *buktighet*

buffer zone *rph* ● zone which accomplishes a smooth transition of neutron properties between two different regions in a reactor core ↦ reactor core **D** *Pufferzone* **F** *zone tampone* **Pl** *strefa buforowa* **Sv** *buffertzon*

build-up factor *nap* ● (for radiation passage through matter:) a ratio of the total value of a certain radiation parameter at a certain point to the contribution to this value from radiation that reach the point without any previous collision ↦ radiation **D** *Zuwachsfaktor; Aufbaufaktor* **F** *facteur d'accumulation* **Pl** *współczynnik narastania* **Sv** *tillväxtfaktor*

built-in reactivity *rph* • excess reactivity in a clean (that is not containing fission products) and cold (at the ambient temperature) nuclear reactor, when all control rods are removed ↦ reactivity; excess reactivity **D** *anfängliche Überschußreaktivität* **F** *réserve de réactivité* **Pl** *reaktywność wbudowana* **Sv** *inbyggd reaktivitet*

bulk boiling ↦*saturated boiling*

burnable absorber *rph* • neutron absorber which is deliberately placed in a reactor core and which is used ("burned") due to neutron absorption; in this way the diminishing reactor reactivity due to fuel burn-up is compensated by the absorber burnup ↦ neutron absorber; reactivity **D** *abbrennbarer Absorber* **F** *absorbeur consommable* **Pl** *wypalający się absorbent* **Sv** *brännbar absorbator*

burnable absorber fuel *nf* • nuclear fuel which contains a burnable absorber ↦ nuclear fuel; burnable absorber; neutron absorber **D** *Brennstoff mit abbrennbarem Absorber* **F** *combustible avec absorbeur consommable* **Pl** *paliwo z wypalającym się absorbentem* **Sv** *BA-bränsle*

burnable poison ↦*burnable absorber*

burner reactor *rty* • nuclear reactor in which the number of created fissile nuclei due to conversion is small and the conversion ratio is less than unity ↦ conversion; conversion ratio **D** *Brenner* **F** *réacteur brûleur* **Pl** *reaktor wypalający* **Sv** *brännreaktor*

burnup *nf* • conversion of atomic nuclei in fuel caused by the neutron irradiation in a nuclear reactor ↦ conversion; specific burnup; FIMA; FIFA **D** *Abbrand* **F** *combustion nucléaire* **Pl** *wypalenie (paliwa)* **Sv** *utbränning*

burnup fraction *rph* • ratio of the current number of atomic nuclei in fuel that is undergoing conversion, to the initial number of nuclei ↦ conversion; burnup; specific burnup; FIMA; FIFA **D** *relativer Abbrand* **F** *taux de combustion; taux d'épuisement* **Pl** *stopień wypalenia* **Sv** *utbränningskvot*

burst can *nf* • fuel element with damaged cladding, through which fission products are leaking ↦ fuel element; cladding; fission product **D** *leckendes Brennelement* **F** *rupture de gaine* **Pl** *uszkodzona koszulka (paliwowa)* **Sv** *läckande bränsleelement*

burst cartridge ↦*burst can*
burst slug ↦*burst can*
BWR ↦*boiling water reactor*

Cc

C ↦ *carbon*

Ca ↦ *calcium*

cadmium *mat* • chemical element denoted Cd, with atomic number $Z=48$, relative atomic mass $A_r=112.411$, density 8.69 g/cm^3, melting point 321 °C, boiling point 767 °C, crustal average abundance 0.15 mg/kg and ocean abundance 1.1×10^{-4} mg/L; *c.* has a significant cross section for neutron absorption ($\sigma_a = 2450$ b for thermal neutrons) which makes it a useful material for control-rod alloys (such as Ag-In-Cd alloy used in PWR control elements) ↦ element; control element **D** *Cadmium* **F** *cadmium* **Pl** *kadm* **Sv** *kadmium*

cadmium ratio *mt* • ratio of neutron intensity observed by a detector with and without a cadmium screen of a specified thickness ↦ cadmium; radiation detector **D** *Cadmiumverhältnis* **F** *rapport cadmique* **Pl** *współczynnik kadmowy* **Sv** *kadmiumkvot*

calandria *rcs* • closed reactor vessel or moderator tank, with penetrating pipes that are designed to separate moderator from coolant, to allow irradiation, or to contain pressure pipes with fuel channels ↦ CANDU reactor; reactor vessel; moderator tank; fuel channel **D** *Kalandriagefäß* **F** *calandre* **Pl** *kalandria; zbiornik rurowy* **Sv** *rörtank*

calcination *wst* • heating up of a substance in air to a temperature below the melting temperature until moisture and volatile substances are removed and the important components transformed into stable oxides ↦ incineration; yellow cake **D** *Kalzinierung* **F** *calcination* **Pl** *prażenie kalcynujące* **Sv** *kalcinering*

calcium *mat* • chemical element denoted Ca, with atomic number $Z=20$, relative atomic mass $A_r=40.078$, density 1.54 g/cm^3, melting point 842 °C, boiling point 1484 °C, crustal average abundance 4.15×10^4 mg/kg and ocean abundance 412 mg/L; *c.* is one of the impurities that are controlled and limited in the primary circuit of a PWR with respect to their impact on crud deposition ↦ element; crud **D** *Calcium* **F** *calcium* **Pl** *wapń* **Sv** *kalcium*

can ↦ *cladding*

canal *rcs* • water-filled space in a decay pond where nuclear fuel and other strongly radioactive parts from a nuclear reactor are placed before further transportation, treatment or storage ↦ decay pond; spent fuel pool **D** *Entladungskanal* **F** *canal; tube de transfert* **Pl** *kanał transportu paliwa* **Sv** *bränsletransportkanal*

CANDU reactor *rty* • (acronym for *CAN*ada *D*euterium *U*ranium:) type of a pressurized heavy-water reactor designed in Canada and used for generating the electric power, employing the natural uranium as a fuel, and the heavy water both as a moderator and coolant; all power reactors of this kind have been built in Canada but have been marketed to other countries such as Argentina, China, India, Pakistan, Romania and South Korea ↦ PHWR; ACR; heavy water **D** *CANDU-Reaktor* **F** *réacteur CANDU* **Pl** *reaktor CANDU* **Sv** *CANDU-reaktor*

canister[1] *wst* • inner container to encapsulate solid radioactive waste, usually in a cylindrical shape; an outer container is used as the shielding ↦ radioactive waste **D** *Abfallbehälter* **F** *conteneur de déchets; fût[1]* **Pl** *kanister na odpady* **Sv** *avfallsbehålare*

canister[2] *wst* • (within waste management:) container for a solid radioactive waste material; the *c.* keeps the material sealed and isolated from the surroundings, whereas the shielding can be provided by an external container ↦ waste management **D** *Abfallbehälter[2]* **F** *cartouche; cartouche de béton* **Pl** *kapsuła na odpady* **Sv** *kapsel[1]*

canning *nf* • procedure to provide cladding for a nuclear fuel ↦ cladding **D** *Einhüllen; Einhülsen* **F** *gainage[1]* **Pl** *koszulkowanie* **Sv** *kapsling*

capacity factor *roc* • (expressed in percent:) ratio of the total electric energy generated over a time period (E_G) to a product of the plant rated power (N_R) and the calendar time (T_{cal}), $CF = 100E_G/(N_R T_{cal})\%$ ↦ gross power; availability factor **D** *Leistungsausnutzung* **F** *taux de charge* **Pl** *współczynnik wykorzystania mocy* **Sv** *utnyttjningsfaktor; energiuttnytjningsfaktor*

capture *nap* • particle-nucleus interaction in which the particle is absorbed by the nucleus to form a compound nucleus in an excited (high-energy) state; this excess energy is next changed by, e.g., emission of gamma radiation, but the particle is kept in the residual nucleus ↦ radiative capture; capture gamma radiation **D** *Einfang* **F** *capture* **Pl** *wychwyt* **Sv** *infångning*

capture cross section *nap* • cross section for a capture reaction ↦ cross section; capture **D** *Einfangsquerschnitt* **F** *section efficace de capture* **Pl** *przekrój czynny na wychwyt* **Sv** *infångningstvärsnitt*

capture gamma radiation *rd* • gamma radiation emitted after the radiative capture ↦ gamma radiation; radiative capture **D** *Einfanggammastrahlung* **F** *rayonnement gamma de capture* **Pl** *promieniowanie gamma powychwytowe* **Sv** *infångningsgammastrålning*

capture-to-fission ratio ↦*alpha ratio*

carbide fuel ↦*ceramic fuel*

carbon *mat* • chemical element denoted C, with atomic number $Z=6$; *c.* occurs in nature in pure forms as diamonds and *graphite* and in impure forms as coal and charcoal; graphite has relative atomic mass $A_r=12.0107$, density 2.267 g/cm³, melting point 4492 °C, sublimation point 3842 °C, crustal average abundance 200 mg/kg and ocean abundance 20 mg/L; graphite is often selected as a moderator and reflector material due to a relatively high microscopic cross section for scattering $(\sigma_s = 4.75$ b) and low for absorption $(\sigma_a = 0.0034$ b) for thermal neutrons ↦ element; moderator **D** *Kohlenstoff* **F** *carbone* **Pl** *węgiel* **Sv** *kol*

Carlson S_N method ↦*discrete ordinates method*

Carnot cycle *th* • closed series of reversible thermodynamic processes proposed by Nicolas Carnot in 1824, consisting of two reversible isothermal processes and two isentropic processes; *C.c.* is the most efficient theoretical cycle for converting a given amount of thermal energy into work in a heat engine; the efficiency of the *C.c.* is given as a ratio of the work performed by the engine to the heat supplied by the high-temperature heat source; it can be shown that the efficiency depends on the temperatures of the heat source T_H and the heat sink T_L as $\eta = 1 - T_L/T_H$ ↦ Brayton cycle; Rankine cycle; reversible process **D** *Carnot-Prozess* **F** *cycle de Carnot* **Pl** *obieg Carnota* **Sv** *Carnot-processen*

cascade *nf* • (in isotope separation:) several separative elements or stages coupled in such a way that the separation effect of a single element or stage is multiplied ↦ isotope separation; separative element; stage **D** *Kaskade* **F** *cascade* **Pl** *kaskada* **Sv** *kaskad*

cascade tails assay *nf* • concentration of one or several isotopes in a cascade tail ↦ cascade; tail **D** *Abstreifkonzentration einer Kaskade* **F** *teneur de rejet d'une cascade* **Pl** *koncentracja resztkowa w kaskadzie* **Sv** *resthalt för en kaskad*

cave ↦ *hot cell*

cell correction factor *rph* • factor introduced to correct calculations of certain reactor parameters resulting from simplification of the reactor cell's shape ↦ reactor cell **D** *Zellkorrekturfaktor* **F** *facteur de correction de cellule* **Pl** *komórkowy współczynnik korekcji* **Sv** *cellkorrektionsfaktor*

Celsius (temperature) scale *th* • (also referred to as the *centigrade scale*, denoted °C:) temperature scale used by the International System of Units SI with the degree Celsius as a derived unit of temperature; a conversion from the *C.t.s.* to the Kelvin scale is: $[K] = [°C] + 273.15$, and to the Fahrenheit scale: $[°F] = [°C] \times \frac{9}{5} + 32$ ↦ degree Celsius; absolute temperature; Fahrenheit (temperature) scale; kelvin **D** *Celsius-Temperaturskala* **F** *échelle de température Celsius* **Pl** *skala (temperatur) Celsjusza* **Sv** *Celsius-skala*

cementation *wst* • binding of radioactive wastes by infusing them in cement ↦ radioactive waste **D** *Zementierung* **F** *cimentation* **Pl** *cementowanie* **Sv** *cementingjutning; cementering*

cent *rph* • (in reactor physics:) one of the units of reactivity, defined as $1/100$-th of a dollar ↦ dollar; reactivity; pcm **D** *Cent* **F** *cent* **Pl** *cent* **Sv** *cent*

centigrade scale ↦ *Celsius (temperature) scale*

centrifugal process *nf* • process of separation of fluid or gaseous mixtures of isotopes using centrifuges ↦ isotope separation **D** *Zentrifugenverfahren* **F** *centrifugation* **Pl** *proces wirówkowy* **Sv** *centrifugprocess*

ceramic fuel *nf* • nuclear fuel consisting of a ceramic material, e.g., an oxide or a carbide; *oxide fuels* $(U,Pu)O_2$ demonstrate very satisfactory dimensional and radiation stability, as well as chemical compatibility with cladding and coolant materials, but have rather low thermal conductivity and low fissile atom density; *carbide fuels* $(U,Pu)C$ have relatively high density and good thermal conductivity, but not very good radiation stability and may cause the cladding carburization ↦ nuclear fuel **D** *keramischer Brennstoff* **F** *combustible céramique* **Pl** *paliwo ceramiczne* **Sv** *kerambränsle; keramiskt bränsle*

cermet fuel *nf* • nuclear fuel consisting of a mixture of a ceramic and a metallic material, which both contain a fissile material; the minimum amount of the required non-heavy metal is normally high, which makes the *c.f.* unattractive for large-scale reactor applications ↦ nuclear fuel **D** *Kermet-Brennstoff* **F** *combustible cermet* **Pl** *paliwo ceramiczno-metalowe; paliwo cermetalowe* **Sv** *kermetbränsle*

chain decay *rdy* • successive radioactive decays, which create a decay chain ↦ radioactive decay; decay chain **D** *Kettenzerfall* **F** *désintégration en chaîne* **Pl** *rozpad łańcuchowy* **Sv** *kedjesönderfall*

Chalk River unidentified deposit ↦*crud*

channeling effect *nap* • increased transmission of radiation in a substance due to sprawled holes or other areas with reduced damping of radiation ↦ radiation **D** *Kanaleffekt* **F** *effet de canalisation* **Pl** *efekt kanałowy* **Sv** *kanalverkan*

channel power *nf* • power generated from a nuclear fuel in a certain cooling channel ↦ nuclear fuel **D** *Kanalleistung* **F** *puissance de canal* **Pl** *moc kanału* **Sv** *kanaleffekt*

charge[1] ↦*fuel charge*

charge[2] *nf* • to introduce nuclear fuel into a reactor core ↦ fuel charge; fuel charging machine **D** *laden* **F** *charger* **Pl** *ładować* **Sv** *ladda*

charger-reader *rdp* • apparatus for reading, and in some cases, for charging of a pen dosimeter ↦ pen dosimeter **D** *Dosimeter-Auflade- und Ablesegerät* **F** *chargeur-lecteur* **Pl** *ładowarka-czytnik (dawkomierza piórowego)* **Sv** *penndosimeteravläsare*

charging pump *rcs* • (in pressurized water reactors:) pump in the chemical and volume control system used for coolant make-up in the primary-coolant system ↦ chemical and volume control system; coolant make-up; primary-coolant system **D** *Ladepumpe* **F** *pompe de charge* **Pl** *pompa napełniająca* **Sv** *laddpump*

chemical and volume control system *rcs* • (abbreviated CVCS:) system connected to the primary-coolant circuit of a pressurized water reactor that consists of three separate sub-systems, which fulfill the following functions: 1) coolant make-up in the primary system, whose volume can change due to controlled leakages, 2) removal of various metallic oxides from the coolant, 3) control (increase or decrease) of the boric acid concentration in the primary circuit; the *c.v.c.s.* can serve as a high-pressure injection system in case of loss-of-coolant accident ↦ primary-coolant circuit; boron control; loss-of-coolant accident **D** *Volumenregelungssystem* **F** *circuit de contrôle chimique et volumétrique* **Pl** *układ uzupełniania i regulacji chemicznej* **Sv** *volymreglersystem*

chemical dosimeter *rdp* • dosimeter containing a chemical substance which undergoes a measurable chemical transformation when irradiated ↦ dosimeter; irradiation **D** *chemisches Dosimeter* **F** *dosimètre chimique* **Pl** *dozymetr chemiczny* **Sv** *kemisk dosimeter*

chemical element ↦*element*

chemical shimming *roc* • reactivity compensation in a nuclear core by using chemical compounds that absorb neutrons; chemical compounds can be dissolved in a liquid coolant, a liquid moderator or any other liquid present in the core ↦ reactivity; absorption control; boron control; fuel control; moderator control; recirculation control; reflector control; spectral shift control **D** *chemisches Trimmen* **F** *compensation chimique* **Pl** *regulacja chemiczna* **Sv** *kemisk styrning*

Cherenkov radiation *rd* • electromagnetic radiation created when charged particles move in a substance with a velocity greater than the radiation phase velocity in the same substance; the *C.r.* can be observed as a weak bluish white glow in pool reactors and in spent fuel pools, where decaying fission products are releasing high-energy electrons (beta radiation) ↦ beta radiation *D Tscherenkow-Strahlung F rayonnement Tcherenkov Pl promieniowanie Czerenkowa Sv Tjerenkovstrålning*

Chernobyl nuclear accident *rs* • nuclear accident that took place in Chernobyl nuclear power plant in Ukraine on 26th of April 1986; due to a number of wrong steps undertaken during an experiment conducted at the plant, the power of the reactor suddenly increased leading to a fire in the graphite moderator and a release of a large amount of radioactive fission products to the atmosphere; the accident was classified as level 7 (the highest one) on the INES scale ↦ INES; Fukushima Daiichi nuclear accident; Three Mile Island nuclear accident *D Nuklearkatastrophe von Tschernobyl F catastrophe nucléaire de Tchernobyl Pl awaria w Czarnobylu Sv Tjernobylolyckan*

CHF ↦ *critical heat flux*

chop and leach *nf* • way of preparation of the spent fuel to fuel reprocessing through chopping fuel assemblies into parts and then selectively dissolving the fuel material through leaching with acid ↦ spent fuel; fuel assembly; fuel reprocessing *D Zerschneiden und Auslaugen F tronçonnage-dissolution Pl cięcie i ługowanie Sv kapning och lakning*

chromium *mat* • chemical element denoted Cr, with atomic number $Z=24$, relative atomic mass $A_r=51.9961$, density 7.15 g/cm^3, melting point 1907 °C, boiling point 2671 °C, crustal average abundance 102 mg/kg and ocean abundance

3×10^{-4} mg/L; isotope ^{51}Cr can be activated with neutrons ↦ element *D Chrom F chrome Pl chrom Sv krom*

Ci ↦ *curie*

CIM ↦ *thermal conductivity*

circumferential ridging *nf* • necking of fuel pellets with ceramic fuel due to strong temperature gradients ↦ fuel pellet; ceramic fuel *D Bambuseffekt F effet bambou Pl efekt bambusowy Sv bambueffekt*

cladding *nf* • sealed casing of a nuclear fuel, which is intended to prevent chemical reactions between the fuel and coolant and to retain radioactive fission products inside the fuel; in light-water reactors, fuel pellets are stacked inside the cladding, which has the form of a tube made of a zirconium alloy ↦ nuclear fuel; fuel pellet; fission product; zirconium; zircaloy *D Brennstoffhülle; Brennstoffhülse F gaine; gaine libre Pl koszulka paliwowa Sv bränslekapsel; kapsel^2*

cladding waste *wst* • radioactive waste that mainly consists of chopped cladding ↦ cladding; radioactive waste *D Hülsenabfall F déchets de dégainage; déchets de gaine; coques Pl odpad koszulek paliwowych Sv kapselavfall*

Clapeyron's equation of state ↦ *ideal gas law*

classified area *rdp* • area in a nuclear installation, in which a radioactive material can be stored or radiation work can be performed; due to radiation protection reasons in the *c.a.*, special rules apply for the access, clothes, eating, etc. ↦ nuclear installation; radioactive material; unclassified area; radiation work *D klassifizierter Bereich F zone classée Pl strefa sklasyfikowana Sv zonindelat område; klassat område*

clean *rph* • (about nuclear reactor:) without fission products and with no induced radioactivity ↦ fission product; induced radioactivity **D** *sauber* **F** *propre* **Pl** *czysty; niezatruty* **Sv** *ren; oförgiftad*

clearance *nf* • (also called *gap*:) space between the fuel pellet and the inner surface of the cladding in a fuel element ↦ fuel pellet; cladding; fuel element **D** *Spalt* **F** *jeu* **Pl** *szczelina* **Sv** *bränslespalt; bränslespel*

cluster control rod *roc* • (for pressurized water reactor:) control element consisting of neutron-absorbing rods connected together on one side and inserted into empty positions in a certain fuel assembly in a reactor core ↦ control element; fuel assembly **D** *Fingersteuerstab* **F** *faisceau de barres de commande* **Pl** *wiązka prętów regulacyjnych* **Sv** *fingerstyrstav*

Co ↦*cobalt*

coarse control element *roc* • control element for coarse adjustment of a reactor core reactivity or for a change of the neutron flux density distribution ↦ control element; neutron flux density **D** *Grobsteuerelement* **F** *élément de réglage grossier* **Pl** *element do regulacji zgrubnej* **Sv** *grovstyrelement*

coarse control member ↦*coarse control element*

coarse control rod *roc* • rod-shaped coarse control element ↦ coarse control element **D** *Grobsteuerstab* **F** *barre de compensation* **Pl** *pręt do regulacji zgrubnej* **Sv** *grovstyrstav*

coastal siting *gnt* • siting of a nuclear installation close to a coast to allow usage of the sea for both cooling and transportation ↦ off-shore siting **D** *Verlegung an der Küste* **F** *établissement sur la côte* **Pl** *lokalizacja przybrzeżna* **Sv** *kustförläggning*

coast-down *roc* • (based on economic considerations:) extension of a length of fuel cycle beyond the nominal one, with a gradual decrease of the reactor power due to loss of reactivity of the reactor core ↦ stretch-out; fuel cycle **D** *Streckbetrieb* **F** *prolongation de cycle* **Pl** *reaktywnościowy wybieg reaktora* **Sv** *utbränningsbetingad effektnedgång*

coated particle *nf* • corn of a fissile or a fertile material with a coating to encapsulate fission products, used in, e.g., the TRISO fuel ↦ fissile material; fertile material; fission product; TRISO **D** *beschichtetes Brennstoffteilchen; beschichtetes Teilchen* **F** *particule enrobée* **Pl** *ziarno paliwa otulone* **Sv** *belagt bränslekorn; belagt korn*

cobalt *mat* • chemical element denoted Co, with atomic number $Z=27$, relative atomic mass $A_r=58.933200$, density 8.86 g/cm^3, melting point 1495 °C, boiling point 2927 °C, crustal average abundance 25 mg/kg and ocean abundance 2×10^{-5} mg/L; c. should be avoided in construction materials since ^{59}Co (with natural abundance 100%) transforms into long-lived (5.3y half-life) radioactive isotope ^{60}Co after irradiation with neutrons ↦ element; induced radioactivity; radioactive half-life **D** *Cobalt* **F** *cobalt* **Pl** *kobalt* **Sv** *kobolt*

Colburn correlation *th* • experimental correlation used to determine the Stanton number as a function of the Reynolds and Prandtl numbers; the correlation is valid for turbulent heat transfer to non-metallic liquids and gases in round tubes and is given as: St $= 0.023$ Re$^{-0.2}$Pr$^{2/3}$, where St is the Stanton number, Re is the Reynolds number and Pr is the Prandtl number; the *C.c.* is valid when $L/D > 60$, Re$> 10^4$ and $0.7 <$ Pr < 160, where L is the distance to the tube entrance and D is the tube diameter ↦ Dittus-Boelter correlation; Prandtl number; Reynolds number; Stanton number **D** *Colburn Gleichung* **F** *corrélation de Colburn* **Pl** *korelacja Colburna* **Sv** *Colburns samband*

cold shutdown *roc* • reactor shutdown with a following significant reduction of the core temperature such that it is much lower than the core temperature during normal operation of the reactor ↦ shutdown; hot standby **D** *kaltes Stillsetzen; kalte Stillegung* **F** *arrêt à froid* **Pl** *zimne wyłączenie* **Sv** *kallavställning*

cold testing *nf* • (in fuel processing:) testing of a method, a process, an apparatus or an instrument in which a strongly radioactive material is replaced with an inactive material with radioactive traces ↦ radioactive material; hot testing; nuclear fuel **D** *kalte Prüfung; inaktive Prüfung* **F** *essai en inactif* **Pl** *zimne testowanie* **Sv** *kallprovning*

cold trap *rcs* • (for liquid sodium:) device to remove impurities (usually sodium oxides) from the circulating sodium, through lowering the temperature to a value at which the impurities are solidified ↦ sodium **D** *Kühlfalle* **F** *piège froid* **Pl** *zimna pułapka* **Sv** *kylfälla*

collapsible cladding *nf* • cladding constructed in such a way that due to the coolant over-pressure it is pressed to, and supported by, the nuclear fuel ↦ cladding; freestanding cladding **D** *Andrückhülle* **F** *gaine non résistante* **Pl** *koszulka niesamonośna* **Sv** *icke fristående kapsel*

collective dose *rdp* • sum of effective dose equivalents for all individuals of a population in a given region or on the whole Earth as a consequence of using certain installations or devices which involve radiation, e.g., nuclear power plants ↦ dose equivalent; effective dose equivalent; detriment; dose com-

mitment **D** *Kollektivdosis; Bevölkerungsdosis* **F** *dose collective* **Pl** *dawka zbiorowa* **Sv** *kollektivdos; kollektivdosekvivalent*

collectron *mt* • (also called self-powered neutron *(SPN) detector*, primary emission neutron activation *(PENA) detector, beta emission detector, Hilborn detector* or *electron emission detector*:) neutron detector in which an electric current is generated due to a beta radiation from a short-lived radioactive nuclide created by a neutron capture; the *c.* is widely applied for in-core usage; the principle of operation is based on measurement of the current corresponding primarily to the beta rays given off by the emitter (a rhodium wire) and flowing to the outer shell, called the collector ↦ radiation detector; beta radiation; nuclide; capture **D** *Kollektron* **F** *collectron* **Pl** *komora β-prądowa; kolektron* **Sv** *betaströmdetektor*

collision density *rph* • number of collisions between neutrons and nuclei per unit volume and time given as $\Sigma_s(E)\phi(E)dE$, where $\Sigma_s(E)$ - macroscopic scattering cross section at energy E, $\phi(E)$ - neutron flux of energy E per unit energy interval and dE - infinitesimal element in energy space ↦ scattering cross section; nucleus; neutron; neutron flux density **D** *Stoßdichte* **F** *densité de collision; densité de chocs* **Pl** *gęstość zderzeń* **Sv** *stöttäthet; kollisionstäthet*

commissioning *roc* • number of steps following the nuclear power plant construction work and assembling, having as a goal to bring the plant to the normal operation at the rated power; the *c.* period can be divided into the following stages: (i) testing of systems and components, (ii) fuel loading into the core and the criticality test of the cold core, (iii) reactor heat-up and testing at a low power, (iv) power increasing, and (v) test operation ↦ reactor start-up **D** *Indienststellung* **F** *démarrage* **Pl** *przekazywanie do eksploatacji* **Sv** *driftsättning*

compound nucleus ↦*resonance energy*

concurrent centrifuge *nf* • centrifuge for isotope separation of gaseous substances in which gas rotates in only one direction ↦ isotope separation **D** *Durchstromcentrifuge* **F** *centrifugeuse à courant parallèle* **Pl** *wirówka współbieżna* **Sv** *medströmscentrifug*

condensation pool *rcs* • (for a boiling water reactor, also called *pressure-suppression pool* or *pressure-suppression chamber*:) chamber filled with water, located in the secondary containment vessel of a boiling water reactor; in general, if all the coolant water vaporizes in a loss-of-coolant accident, about 90% of the resulting steam should condense in the *c.p.* to prevent the containment pressure from exceeding specified limits ↦ pressure-suppression; drywell; wetwell; loss-of-coolant accident **D** *Kondensationsbecken* **F** *piscine de condensation* **Pl** *basen kondensacyjny* **Sv** *kondensationsbassäng*

condenser *rcs* • heat exchanger in a steam-generating loop in which the steam leaving a turbine is condensed to the liquid phase; the *c.* has two functions: (i) it transfers heat from the steam-generating loop to the ultimate heat sink and (ii) it provides very low pressure ("vacuum") at the turbine exit, which improves the turbine thermodynamic efficiency ↦ condensing power **D** *Kondensator* **F** *condensateur* **Pl** *skraplacz* **Sv** *kondensor*

condensing power *th* • electric power provided by a turbo-generator, in which the exit steam from the tur-

bine condenses in a condenser cooled with cold water, without further use of the exit steam enthalpy ↦ condenser **D** *Kondensationsleistung* **F** *puissance à l'état condensé* **Pl** *moc kondensacyjna* **Sv** *kondenseffekt*

condition for initiating emergency action *rs* • operational situation or past event which causes the emergency organization to be informed or set in function to an appropriate extent ↦ emergency preparedness; emergency planning zone; reference level for emergency action; INES **D** *Kriterium für die Einleitung von Notmaßnahmen; Kriterium für Katastrophenalarm* **F** *critère d'arrêt d'urgence* **Pl** *kryterium wszczęcia postępowania awaryjnego* **Sv** *larmkriterium*

conductivity integral ↦*thermal conductivity*

conductivity integral to melt ↦*thermal conductivity*

configuration control *roc* • reactor control through change of a position of nuclear fuel, reflector, coolant or moderator ↦ reactor control[2]; reflector; coolant; moderator **D** *Konfigurationssteuerung* **F** *commande par configuration* **Pl** *sterowanie (reaktora) układem rdzenia* **Sv** *konfigurationsstyrning*

container ↦*canister[1]*

containment[1] *rcs* • sealed structure surrounding the reactor coolant system that includes a thick barrier of reinforced concrete, to withstand high pressure that may occur during an accident, and a steel liner, to prevent leakage of radioactive substances to the environment; the *c.* also protects the reactor from external events; an *inerted containment* is filled with a nitrogen gas at atmospheric pressure to eliminate the risk of hydrogen explosion in accident situations ↦ reactor pressure vessel; primary-coolant circuit;

accident **D** *Reaktorsicherheitshülle; Sicherheitshülle* **F** *enceinte de confinement (d'un réacteur)* **Pl** *obudowa bezpieczeństwa[1]; budynek bezpieczeństwa* **Sv** *reaktorinneslutning; inneslutning[1]; inneslutningsskal*

containment[2] *wst* • (at waste management:) placement of a radioactive waste in such a way that its spreading to the surroundings is prevented ↦ waste management; radioactive waste **D** *Einschließung* **F** *confinement* **Pl** *obudowa bezpieczeństwa[2]* **Sv** *inneslutning[2]*

contamination ↦*radioactive contamination*

control element *roc* • movable reactor component that directly influences the reactivity and which is used for reactor control ↦ reactor control[1,2]; reactivity **D** *Steuerelement* **F** *élément de commande* **Pl** *organ regulacyjny* **Sv** *styrelement*

controlled area *rs* • defined area in which specific protection measures and safety provisions are, or could be, required for controlling normal exposures or preventing the spread of contamination during normal working conditions, and preventing or limiting the extent of potential exposures; the *c.a.* is often within a supervised area, but need not be ↦ site area; supervised area **D** *kontrollierte Zone* **F** *zone contrôlée* **Pl** *obszar pod kontrolą* **Sv** *kontrollerat område*

controlled tipping *wst* • (also called *landfill, landfilling* or *deposition (of waste)*:) placing of radioactive wastes within a restricted area ↦ waste management; ultimate waste disposal **D** *Abfallagerung; Endlagerung[1]* **F** *stockage des déchets; mise en décharge* **Pl** *składowanie odpadów* **Sv** *avfallsdeponiering*

control margin *roc* • possible variation region of an operation parameter with influence on the reactor con-

trol, such as a coolant flow or a position of a control rod ↦ reactor control[2]; coolant; control rod **D** *Steuerbereich* **F** *marge de réglage* **Pl** *margines regulacji* **Sv** *reglermarginal*

control member ↦*control element*

control rod *roc* • control element of a nuclear reactor having the form of a rod with a circular (PWR) or a cruciform (BWR) cross section; in PWRs the control material commonly used is an alloy of 80% (by weight) silver, 15% indium and 5% cadmium; in BWRs the preferred control material is boron carbide (B_4C); this material is also preferred in gas-cooled reactors and in fast breeder reactors due to its good performance at high temperatures ↦ shimming; boron control; reactivity **D** *Steuerstab* **F** *barre de commande; grappe de commande* **Pl** *pręt regulacyjny* **Sv** *styrstav*

control-rod configuration *roc* • specific position of control rods in a nuclear reactor core; the *c.-r.c.* can be the same during several days or weeks, which is, in particular, valid for BWRs, where the reactor power is controlled through changes of the circulation flow through the core ↦ circulation control; control rod **D** *Steuerstabkonfiguration* **F** *configuration de barre de commande* **Pl** *konfiguracja prętów regulacyjnych* **Sv** *styrstavsmönster*

control-rod ejection *roc* • (in pressurized water reactors:) hypothetical accident during which the housing of a control rod drive system is destroyed and the control rod is ejected from the core due to pressure forces; this type of accident is considered in the safety analysis reports and it corresponds to a reactivity insertion causing a prompt power excursion; due to the *c.-r.e.*, the radial power distribution in its vicinity is significantly deformed; the power excursion is limited by the Doppler effect and terminated by the reactor trip ↦ Doppler effect; reactor trip **D - F - Pl** *wystrzelenie pręta regulacyjnego* **Sv** *styrstavsejektion*

control-rod gap *rcs* ● space between fuel assemblies in a reactor core intended for control rods ↦ narrow water gap **D** *Steuerstabkanal* **F** *domaine de manoeuvre de grappes; fente de barre de commande* **Pl** *szczelina pręta regulacyjnego* **Sv** *styrstavsspalt; bredspalt*

control-rod group *roc* ● group of control rods operated simultaneously during reactor trip or during change of their configuration ↦ control-rod configuration; reactor trip **D** *Steuerstabgruppe; Steuerstabbank* **F** *groupe de barres de commande* **Pl** *zespół prętów regulacyjnych* **Sv** *styrstavsgrupp*

control-rod manoeuvring *roc* ● displacement of control rods using actuators ↦ control rod **D** *Steuerstabbewegung* **F** *mouvement d'une barre de commande; manoeuvre des grappes de contrôle* **Pl** *manewrowanie prętem regulacyjnym* **Sv** *styrstavsmanövrering; styrstavskörning*

control-rod position indicator *roc* ● indication of the position of control rods in a nuclear reactor core ↦ control rod **D** *Positionsanzeige der Steurstäbe* **F** *indication de position de la barre de commande* **Pl** *wskaźnik położenia pręta regulacyjnego* **Sv** *styrstavsindikering*

control-rod sequence *roc* ● series of control rod configurations following after each other ↦ control-rod configuration **D** *Steuerstabfolge* **F** *série de configurations de barres de commande* **Pl** *sekwencja prętów regulacyjnych* **Sv** *styrstavssekvens*

control-rod worth *roc* ● reactivity change resulting from a full insertion of a totally withdrawn control rod into a critical reactor under spe-

cific conditions ↦ control rod; reactivity **D** *Reaktivitätsäquivalent des Steuerstabes* **F** *efficacité d'une barre de commande* **Pl** *wartość pręta regulacyjnego* **Sv** *styrstavsvärde*

control room *rcs* ● compartment connected to the reactor building, placed between the reactor building and the turbine building, from which the nuclear power plant is operated ↦ control-room personnel **D** *Leitwarte* **F** *salle de commande* **Pl** *nastawnia* **Sv** *kontrollrum*

control-room personnel *roc* ● personnel operating the reactor from the control room ↦ control room **D** *Personal der Leitwarte* **F** *personnel de conduite* **Pl** *personel nastawni* **Sv** *kontrollrumspersonal*

convection *th* ● heat transfer mode in a fluid, when heat is transported by its bulk motion; this type of heat transfer is a basis of the cooling of nuclear fuel elements in reactor cores, in which the heat generated due to nuclear fission is transferred to the coolant; there are two fundamental types of the *c.*: (i) natural convection, and (ii) forced convection; when both types of convection co-exist, the resulting heat transfer mode is referred to as *mixed-convection* ↦ natural convection; forced convection; heat transfer coefficient **D** *Konvektion* **F** *convection* **Pl** *konwekcja* **Sv** *konvektion*

convective heat transfer ↦*convection*

conventional flux density *rph* ● fictive neutron flux density equal to a product of the total number of neutrons per unit volume and neutron velocity 2200 m/s, corresponding to neutron energy $E_0 = k_B T = 0.025$ eV, where k_B is the Boltzmann constant and T is the temperature ↦ neutron flux density; Boltzmann constant **D** *konventionelle Flußdichte; 2200-*

*m/s-Flußdichte **F** densité de flux convention-*
*nelle **Pl** konwencjonalna gęstość strumienia*
*neutronów **Sv** konventionell flödestäthet*

conversion *rph; nf* • **1.** transmutation of a fertile material into a nuclear
fuel, e.g., ^{238}U into ^{239}Pu or ^{232}Th
into ^{233}U; **2.** chemical process which
in nuclear engineering refers to transformation of natural or enriched uranium contained in UF_4 or UF_6 into
UO_2 ↦ transmutation; fertile material; enriched uranium **D** *Konversion* **F** *conversion*
Pl *konwersja* **Sv** *konversion; konvertering*

conversion ratio *rph* • (abbreviated
CR:) ratio of the number of fissile nuclei produced to the number of fissile
nuclei destroyed in a reactor; in thermal reactors based on uranium fuel,
the *c.r.* is less than unity; the ratio
is called a breeding ratio when it is
greater than unity and breeding occurs; for a given time during reactor
operation the *c.r.* is defined as a ratio
of the rate of formation of fissile nuclei
to the rate of destruction of fissile nuclei; since these rates vary during reactor operation, the *c.r.* varies as well;
for thermal reactors with natural or
slightly enriched uranium as fuel, the
initial value of *c.r.* can be found as
$CR = N^{238}\sigma_c^{238} / \left(N^{235}\sigma_a^{235}\right) + r$ with
$r = \varepsilon P_1(1-p)\eta^{235}$, where σ_c^{238} is the
capture cross section of ^{238}U for ther-
mal neutrons, N^{238} is the concentration of ^{238}U nuclei, N^{235} is the concentration of ^{235}U nuclei, σ_a^{235} is the
absorption cross section of ^{235}U for
thermal neutrons, ε is the fast-fission
factor, η^{235} is the number of fast neutrons produced per neutron absorbed
in ^{235}U, P_1 is the non-leakage probability in slowing down into the resonance energy region and p is the resonance escape probability; in commercial light-water reactors the initial *c.r.*

is approximately 0.6; taking into account losses of plutonium due to the
neutron capture, one-third of the heat
produced in such reactors results from
the fission of ^{239}Pu and ^{241}Pu ↦ fissile
material; fertile material; breeding ratio **D**
Konversionsverhältnis **F** *rapport de conver-*
sion **Pl** *współczynnik konwersji* **Sv** *konver-*
sionsförhållande

converter reactor *rty* • nuclear reactor in which the rate of formation
of fissile nuclei is less than the rate
of their destruction ↦ breeder reactor
D *Konverter* **F** *réacteur convertisseur* **Pl**
reaktor-konwertor **Sv** *konversionsreaktor*

coolant *th* • fluid circulated through
the reactor core to remove the heat
generated in it due to fissions; several
different types of *c.* have been used
so far; among the most widespread
ones are light water, liquid sodium
(or sodium-potassium alloy), carbon
dioxide, helium and certain organic
compounds ↦ coolant channel; coolant
make-up **D** *Kühlmittel* **F** *caloporteur;*
réfrigérant; fluide de refroidissement **Pl**
chłodziwo **Sv** *kylmedel*

coolant channel *th* • channel in a
reactor core through which coolant
flows in the fuel lattice; this can be an
actual channel in the fuel assembly, as
in high-temperature gas-cooled reactors, or an equivalent channel associated with a single fuel rod, as in light-
water reactors ↦ fuel channel **D** *Kühl-*
lkanal **F** *canal de refroidissement* **Pl** *kanał*
chłodzący **Sv** *kylkanal*

coolant circuit *th* • system used
to transport heat by a circulating
coolant ↦ coolant **D** *Kühlkreislauf; Küh-*
lkreis **F** *circuit de refroidissement* **Pl** *obieg*
chłodzący **Sv** *kylkrets*

coolant make-up *th* • refilling
coolant into the primary-coolant circuit of a nuclear reactor to assure

proper conditions for effective cooling of the core ↦ coolant; primary-coolant circuit **D** *Kühlmittelaufbereitung* **F** *circuit d'appoint en réfrigérant* **Pl** *uzupełnianie chłodziwa* **Sv** *spädmatning*

cooling pond *th* • (also called *fuel-cooling installation:*) system for cooling of the used fuel ↦ decay pond **D** *Abklingbecken*[1] **F** *installation de refroidissement du combustible; piscine de désactivation*[1] **Pl** *staw chłodzący* **Sv** *bränsleavklingningssystem*

cooling tower *th* • closed recirculation system for cooling of the condenser water; two types of the *c.t.* exist: (i) the wet *c.t.*, where the condenser discharge water flows down as droplets and most of the cooling is due to vaporization causing substantial loss of water, and (ii) the dry *c.t.*, in which the heat is removed mainly by convection to the ambient air and there is no loss of water by evaporation ↦ condenser **D** *Kühlturm* **F** *tour de refroidissement; réfrigérant atmosphérique* **Pl** *chłodnia kominowa* **Sv** *kyltorn*

copper *mat* • chemical element denoted Cu, with atomic number $Z=29$, relative atomic mass $A_r=63.546$, density 8.933 g/cm^3, melting point 1084.62 °C, boiling point 2562 °C, crustal average abundance 60 mg/kg and ocean abundance 2.5×10^{-4} mg/L; isotope ^{63}Cu (with natural abundance 69%), due to (n,γ) reaction, is transformed into radioactive isotope ^{64}Cu with radioactive half-life 12.8 h and maximum gamma-ray energy 1.35 MeV ↦ element; induced radioactivity; radioactive half-life **D** *Kupfer* **F** *cuivre* **Pl** *miedź* **Sv** *koppar*

core accident *rs* • destruction of a reactor core due to the core overheating or the reactor excursion; the term *severe accident* is used when

the *c.a.* results in a significant structural degradation of the reactor core, including the core melt ↦ core damage; core overheating; reactor excursion; core melt **D** *Kernschaden*[1] **F** *accident du coeur* **Pl** *awaria rdzenia* **Sv** *härdhaveri*

core analysis *th* • calculation of the multiplication factor, the power distribution and other parameters of a nuclear reactor core ↦ multiplication factor **D** *Kernberechnung* **F** *analyse du coeur; calculus physiques du coeur* **Pl** *analiza rdzenia* **Sv** *härdberäkning; härdanalys*

core baffle *th* • plate construction around the reactor core in a pressurized water reactor intended to guide the cooling water ↦ pressurized water reactor; core barrel **D** *Kernbehälter*[1] **F** *cloisonnement du coeur* **Pl** *przegroda rdzenia* **Sv** *härdbaffel*

core barrel ↦*moderator tank*

core catcher *rcs* • spreading compartment below the reactor vessel whose primary function is to collect, cool down and stabilize the molten core debris (also called the *corium*) in the event of a severe accident ↦ core melt; core debris; core accident; reactor vessel **D - F - Pl** *chwytacz rdzenia* **Sv** *härdfångare*

core damage *rs* • damage in a reactor core due to the core overheating or reactor excursion ↦ core accident; core overheating; reactor excursion **D** *Kernschaden*[2] **F** *dégâts causés au coeur (pl)* **Pl** *uszkodzenie rdzenia* **Sv** *härdskada*

core debris *rs* • debris resulting from a destroyed reactor core or from solidified remains of a core melt ↦ core melt **D** *Kernschüttbett* **F** *débris du coeur (pl)* **Pl** *szczątki rdzenia (pl)* **Sv** *härdfragment*

core grid *rcs* • grid-like structure located in the upper part of a reactor core whose main purpose is to support fuel assemblies, neutron flux detectors and start-up neutron sources ↦ fuel assembly; start-up neutron source **D** *Kerngitter* **F** *sommier (du coeur)* **Pl** *szkielet rdzenia* **Sv** *härdgaller*

core melt *rs* • severe damage of the reactor core during which the core is partially or completely melted and relocated in the reactor vessel ↦ core catcher; reactor vessel; core damage; core debris **D** *Kernschmelze*[1] **F** *fusion du coeur*[1] **Pl** *stopienie rdzenia* **Sv** *härdsmälta*

core meltdown *rs* • process that leads to a core melt ↦ core melt **D** *Kernschmelze*[2] **F** *fusion du coeur*[2] **Pl** *topienie się rdzenia* **Sv** *härdsmältning*

core overheating *rs* • exceeding of the highest allowed temperature of the fuel in the reactor core ↦ core melt **D** *Kernüberhitzung* **F** *surchauffe du cœur* **Pl** *przegrzanie rdzenia* **Sv** *härdöverhettning*

core spray system *rcs* • one of the emergency core-cooling subsystems in boiling water reactors designed to be effective over the whole range of pressure during a postulated design-basis loss-of-coolant accident; the *c.s.s.* pumps water from the pressure-suppression pool through ring spargers, mounted inside the core shroud in the space between the top of the core and the steam separator base; the core spray ring spargers are provided with spray nozzles for the injection of cooling water ↦ emergency core cooling system; pressure-suppression system **D** *Kernsprühsystem* **F** *système d'aspersion du coeur* **Pl** *układ zraszania rdzenia* **Sv** *härdstrilsystem*

core structure *rcs* • construction placed in a reactor vessel and usually composed of several parts, such as, e.g., a moderator tank and a core grid ↦ reactor vessel; moderator tank; core grid **D** *Kernbehälter*[2] **F** *support du coeur*

Pl *podpora rdzenia* **Sv** *härdstomme; reaktorhärdstomme*

corium ↦ *core catcher*

count[1] *mt* • numerical value of the output quantity from a scaler ↦ scaler **D** *Zählung*[1] **F** *coup*[1] **Pl** *zliczenie*[1] **Sv** *pulstal*

count[2] *mt* • impulse produced by a counter ↦ counter **D** *Zählung*[2] **F** *coup*[2] **Pl** *zliczenie*[2] **Sv** *räknarpuls*

counter *mt* • radiation detector that generates impulses ↦ scaler **D** *Zähler*[1] **F** *compteur* **Pl** *licznik* **Sv** *räknare*

counter-current centrifuge *nf* • centrifuge for the isotope separation of gaseous substances in which the gas rotates in opposite directions at the center and at the periphery, causing the isotope exchange between the two streams ↦ isotope separation; concurrent centrifuge **D** *Gegenstromcentrifuge* **F** *centrifugeuse à contre-courant* **Pl** *wirówka przeciwprądowa* **Sv** *motströmscentrifug*

counter range *mt* • power range of a nuclear reactor in which the neutron flux density is too low to be measured with ordinary instrumentation but instead it has to be measured with a particle-counting instrument ↦ isotope separation; concurrent centrifuge **D** *Zählerbereich* **F** *domaine de comptage* **Pl** *zakres licznikowy* **Sv** *räknarområde*

counter tube *mt* • radiation detector consisting of a gas-filled tube for which the gas amplification coefficient is much higher than unity and in which the single, primary ionization processes generate discrete electric pulses ↦ radiation detector **D** *Zählrohr* **F** *tube compteur* **Pl** *licznik (promieniowania)* **Sv** *räknerör*

counting rate *mt* • number of counts per unit time; typical units of the *c.r.*, which are not included in the SI units, are cps (counts per second) and cpm (counts per minute); the *c.r.* does not universally convert to the dose rate and any conversions are instrument-specific ↦ count; dose rate *D* Zählrate *F* taux de comptage *Pl* szybkość zliczania *Sv* pulsrat

counting-rate meter *mt* • device that continuously provides a mean value of counts per unit time ↦ count *D* Zählratenmesser *F* ictomètre *Pl* miernik szybkości zliczania *Sv* pulsratsmätare

CPR ↦critical power ratio

Cr ↦chromium

CR ↦conversion ratio

critical *rph* • (referring to a nuclear reactor:) such reactor status in which a stationary chain reaction is taking place, the effective multiplication factor equals unity and the reactor power is constant; these conditions correspond to the normal operation conditions of a nuclear reactor ↦ effective multiplication factor; supercritical; subcritical; prompt critical *D* kritisch *F* critique *Pl* krytyczny *Sv* kritisk

critical assembly *nf* • assembly which is able to attain criticality and that can be used for criticality investigations ↦ assembly; critical; critical volume; critical mass *D* kritische Anordnung *F* assemblage critique *Pl* zestaw krytyczny *Sv* kritisk uppställning

critical equation *rph* • equation which contains reactor parameters and which must be satisfied for a reactor to be critical; in the one-group diffusion approximation, the *c.e.* describing a critical bare reactor is as follows, $k_\infty/(1 + L^2 B_m^2) = 1$, where k_∞ is the infinite multiplication factor, L is the neutron diffusion length, and B_m^2 is the material buckling ↦ infinite multiplication factor; diffusion length; material buckling; critical; critical volume; critical mass *D* kritische Gleichung *F* équa-

tion critique *Pl* równanie krytyczne *Sv* kriticitetsekvation

critical experiment *mt* • experiment with a device that contains fissile material, during which conditions can be gradually changed until the device will become critical ↦ fissile material; critical *D* kritisches Experiment *F* expérience critique *Pl* doświadczenie krytyczne *Sv* kriticitetsförsök

critical flow *th* • maximum mass flow rate through a nozzle when flow at the nozzle throat is sonic; the *c.f.* will occur for coolant discharge during a loss-of-coolant accident in a light-water reactor ↦ loss-of-coolant accident *D* kritische Ausströmung *F* débit critique *Pl* wypływ krytyczny *Sv* kritisk utströmning

critical group *rdp* • (for given radiation source:) this fraction of the population which creates a relatively homogeneous group and which receives the highest effective dose equivalent from the source ↦ radiation source; effective dose equivalent *D* kritische Bevölkerungsgruppe *F* population critique; groupe critique *Pl* populacja krytyczna *Sv* kritisk grupp

critical heat flux *th* • (abbreviated *CHF*:) heat flux at which a dramatic deterioration of heat transfer between a solid wall and a fluid occurs due to transition from the liquid-phase cooling to the vapour-phase cooling; two types of the *c.h.f.* are, in general, distinguished: (i) the departure from nucleate boiling, in which the nucleate boiling heat transfer transits into the film boiling heat transfer; (ii) the dry-out, in which the liquid film in the annular two-phase flow dries out and the convective boiling heat transfer transits into the post-dryout heat transfer ↦ departure from nucleate boiling; dry-out; film boiling; annular flow *D* kritische

Wärmestromdichte **F** *flux critique* **Pl** *krytyczny strumień ciepła* **Sv** *kritisk yteffekt; kritisk värmeflödestäthet*

criticality *rph* • condition at which a system which contains a fissile material is critical ↦ critical; fissile material **D** *Kritikalität* **F** *criticité* **Pl** *krytyczność* **Sv** *kriticitet*

criticality accident *rs* • accident caused by a fissile material, which unintentionally becomes critical; the material can be located in a nuclear reactor or, e.g., may be a subject to chemical processing ↦ critical; fissile material **D** *Kritikalitätsunfall* **F** *accident de criticité* **Pl** *awaria krytycznościowa* **Sv** *kriticitetsolycka*

criticality point *roc* • (during movement of control rods:) control rod position at which the reactor becomes critical; similar definition can be formulated if the reactor is made critical in a different way ↦ critical; control rod **D** *Kritikalitätspunkt* **F** *point de criticité* **Pl** *punkt krytycznościowy* **Sv** *kriticitetspunkt*

criticality transition *rph* • (in a multiplying system:) transition from the subcritical to supercritical system in which the multiplication factor increases from a sub-unity to greater than unity value ↦ multiplying; subcritical; supercritical; multiplication factor **D** *Kritikalitätspassage* **F** *passage du seuil de criticité; passage de criticité* **Pl** *przejście krytycznościowe* **Sv** *kriticitetspassage*

critical mass *rph* • minimum mass of a fissile material that is capable of sustaining a nuclear chain reaction; the *c.m.* of a reactor material depends on a wide range of conditions such as the configuration and the composition of the material; thus the *c.m.* of ^{235}U may range from less than 1 kg for a system consisting of 90% enriched uranium salt dissolved in water

to 200 kg of ^{235}U present in 30 000 kg of natural uranium embedded in a matrix of graphite; natural uranium alone can never be critical since too many fission neutrons are lost in non-fission reactions ↦ critical size; critical volume; nuclear chain-reaction **D** *kritische Masse* **F** *masse critique* **Pl** *masa krytyczna* **Sv** *kritisk massa*

critical organ *rdp* • this organ in a biological system which is at risk of the most serious consequences in case of exposure to ionizing radiation due to, e.g., an uptake of a certain radioactive nuclide ↦ ionizing radiation; nuclide **D** *kritisches Organ; strahlungsempfindliches Organ* **F** *organe critique* **Pl** *organ krytyczny* **Sv** *kritiskt organ*

critical pathway *rdp* • the dominant chain of processes in nature through which a certain radioactive nuclide reaches the critical group ↦ radionuclide; critical group; critical organ **D** *kritischer Pfad* **F** *voie critique* **Pl** *ścieżka krytyczna; droga krytyczna* **Sv** *kritisk bestrålningsväg*

critical point *th* • end point of a phase equilibrium curve; the most prominent example is the liquid-vapor critical point, the end point of the pressure-temperature curve that designates conditions under which a liquid and its vapour can coexist ↦ critical fluid **D** *kritische Punkt* **F** *point critique* **Pl** *punkt krytyczny* **Sv** *kritisk punkt*

critical power ↦*critical power ratio*

critical power ratio *th* • (abbreviated *CPR:*) ratio of the *critical power* of a fuel assembly q_{cr} (that is, such a thermal power at which the critical heat flux condition for any rod in the assembly will take place) to the actual thermal power of the assembly q_a: $CPR = q_{cr}/q_a$ ↦ critical heat flux **D** *kritisches Leistungsverhältnis* **F** *rapport* *des puissances critiques* **Pl** *stosunek mocy krytycznej* **Sv** *torrkokningskvot*

critical size *rph* ● minimum size of a fissile material that is capable of sustaining a nuclear chain reaction; the *c.s.* of a reactor material depends on a wide range of conditions such as the configuration and the composition of the material ↦ critical mass; critical volume; nuclear chain-reaction **D** *kritische Größe* **F** *taille critique* **Pl** *rozmiar krytyczny* **Sv** *kritisk storlek*

critical volume *rph* ● minimum volume of a fissile material that is capable of sustaining a nuclear chain reaction; the *c.v.* of a reactor material depends on a wide range of conditions such as the configuration and the composition of the material ↦ critical mass; critical size; nuclear chain-reaction **D** *kritisches Volumen* **F** *volume critique* **Pl** *objętość krytyczna* **Sv** *kritisk volym*

cross section *xr* ● measure of the probability of a particular neutron-nucleus interaction; the *c.s.* of a particular reaction that applies to a single nucleus is called the microscopic *c.s.*, whereas the total cross section of the nuclei per unit volume is called the macroscopic *c.s.* ↦ barn; microscopic cross section; macroscopic cross section **D** *Wirkungsquerschnitt* **F** *section efficace* **Pl** *przekrój czynny* **Sv** *tvärsnitt*

crud *mat* ● (acronym for *Chalk River Unidentified Deposit:*) corrosion products in light-water reactors that cover the fuel element surface and other surfaces that are in contact with the coolant; the *c.* deposition on fuel elements can lead to the *Axial Offset Anomaly* (AOA) phenomenon, also referred to as the *Crud-Induced Power Shift* (CIPS) phenomenon, when a reduction in neutron flux on the upper spans of fuel assemblies undergo-ing sub-cooled nucleate boiling occurs due to boron concentration in corrosion product deposits ↦ corrosion **D** *Ablagerung von Korrosionsprodukten* **F** *impureté* **Pl** *osad produktów korozji* **Sv** *crud*

crud-induced power shift ↦*crud*

Cu ↦*copper*

curie *rd* ● (abbreviated Ci:) older unit of the activity, equivalent to $3.7 \cdot 10^{10}$ Bq; 1 Ci is the approximate activity of one gram of the radium isotope ^{226}Ra ↦ activity; becquerel; radium **D** *Curie* **F** *curie* **Pl** *kiur* **Sv** *curie*

curium *rd* ● transuranic radioactive chemical element belonging to the actinide series, denoted Cm, with atomic number $Z = 96$, atomic mass of the most stable isotope $A = 247$, density 13.51 g/cm^3, melting point 1340 °C and boiling point 3110 °C; *c.* is created in small quantities in nuclear reactors through irradiation of uranium and plutonium by neutrons ↦ actinide; synthetic element **D** *Curium* **F** *curium* **Pl** *kiur* **Sv** *curium*

cut *nf* ● ratio of the flow of enriched material from a separative element and the feed flow to the element ↦ enriched material **D** *Aufteilungsverhältnis* **F** *coefficient de partage* **Pl** *frakcja* **Sv** *delning*[1]

cycle efficiency *th* ● ratio of the useful work obtained from a thermodynamic cycle to the thermal energy supplied to the cycle ↦ Carnot cycle; Rankine cycle; Brayton cycle **D** *Wirkungsgrad* **F** *rendement thermodynamique* **Pl** *sprawność obiegu (termodynamicznego)* **Sv** *cykelns verkningsgrad*

cylinder function *mth* ● solution of Bessel's differential equation ↦ Bessel's differential equation; Bessel functions of the first kind; Bessel functions of the second kind; Bessel functions of the third kind **D** - **F** - **Pl** *funkcja cylindryczna* **Sv** *cylinderfunktion*

Dd

d ↦*deuteron*

D ↦*deuterium*

daily load cycling *roc* • adjusting the rector power to the daily change of the demand for electricity ↦ base-load operation; load following **D** *Tagesregelung; Tagregelung; täglicher Lastzyklus* **F** *réglage quotidien de la puissance; réglage quotidien* **Pl** *dzienna regulacja mocy* **Sv** *dygnsreglering*

dalton ↦*atomic mass unit*

Dancoff correction *rph* • (also called *Dancoff factor* or *Dancoff-Ginsberg factor*:) correction added to a resonance integral for a single fuel rod to obtain a resonance integral for the rod in the presence of other similar rods; in closely packed lattices some of the resonance neutrons are intercepted by adjacent fuel rods, which effectively reduces the resonance integral; this effect depends on the spacing and radius of fuel rods and is taken into account by the *D.c.* ↦ resonance integral; fuel rod **D** *Dancoff-Korrektur* **F** *correction de Dancoff* **Pl** *poprawka Dancoffa* **Sv** *Dancoff-korrektion*

Dancoff factor ↦*Dancoff correction*

Dancoff-Ginsberg factor ↦*Dancoff correction*

Darcy friction factor *th* • (denoted λ:) friction factor for a single-phase fluid flow in a channel defined as $\lambda = 4\tau_w/(\frac{1}{2}\rho U^2)$, where τ_w - wall shear stress, ρ - fluid density, U -

mean fluid velocity in the channel cross-section ↦ Fanning friction factor **D** *Rohrreibungszahl* **F** *coefficient de perte de charge de Darcy* **Pl** *współczynnik oporu Darcy-Weisbacha* **Sv** *Darcy-Weisbachs friktionsfaktor*

DBA ↦*design-basis accident*

DBE ↦*design-basis earthquake*

decanning *nf* • (also called *decladding*:) removal of the can from a fuel element ↦ can; fuel element **D** *Enthülsen* **F** *pelage; dégainage* **Pl** *usuwanie koszulek z prętów paliwowych* **Sv** *avkapsling*

decay *rdy* • reduction of the activity through radioactive decay ↦ activity; radioactive decay **D** *Abklingen* **F** *désactivation* **Pl** *zanik; rozpad* **Sv** *avklingning*

decay chain *rdy* • (also called *decay series*:) chain of nuclides formed in subsequent radioactive decays when each nuclide transforms into another nuclide, until a stable nuclide is created; there are three natural *d.c.*: (i) uranium-radium chain, where the starting nuclide is ^{238}U and the final stable nuclide is ^{206}Pb, (ii) thorium chain, in which the initial nuclide is ^{232}Th and the final stable nuclide is ^{208}Pb, (iii) uranium-actinide chain in which ^{235}U is the initial nuclide and ^{207}Pb is the final stable nuclide ↦ nuclide; stable nuclide; radioactive decay; uranium; radium; thorium; actinide **D** *Zerfallsreihe* **F** *chaîne de désintégration* **Pl** *szereg*

---

promieniotwórczy **Sv** *sönderfallskedja; sönderfallsserie*

decay constant *rdy* • (for a certain radioactive nuclide, also called the *disintegration constant*, denoted λ:) quantity equal to the probability that one nucleus will decay during a unit time; the *d.c.* is related to the mean lifetime τ and the radioactive half-life of the nuclide $T_{1/2}$ as follows: $\lambda = 1/\tau = \ln 2/T_{1/2} \approx 0.693/T_{1/2}$ $\mapsto$ nuclear disintegration; radioactive decay; radioactive half-life **D** *Zerfallskonstante* **F** *constante de désintégration* **Pl** *stała rozpadu* **Sv** *sönderfallskonstant*

decay curve *rdy* • curve which describes the activity of a radioactive material, or any of its components, as a function of time $\mapsto$ activity; radioactive material **D** *Zerfallskurve* **F** *courbe de décroissance; courbe d'activité* **Pl** *krzywa rozpadu* **Sv** *sönderfallskurva; avklingningskurva*

decay heat *th* • energy of beta particles and gamma photons of all the individual fission products, which is deposited as heat in a reactor core $\mapsto$ decay power; residual power **D** *Nachwärme* **F** *chaleur résiduelle* **Pl** *ciepło rozpadu; ciepło powyłączeniowe* **Sv** *sönderfallsvärme*

decay-heat removal $\mapsto$ residual heat removal

decay pond *rcs* • water pool in which a spent fuel or another radioactive material is stored and cools down when its radioactivity drops down $\mapsto$ spent fuel pool **D** *Abklingbecken*[2] **F** *piscine de désactivation*[2] **Pl** *staw dezaktywacji* **Sv** *avklingningsbassäng*

decay power *th* • thermal power of a reactor after shutdown, resulting from the radioactive decay of fission products in the core; the *d.p.* can be found from the following relationship $q(t) \approx 0.065 q_0 [(t - t_0)^{-0.2} - t^{-0.2}]$ where q_0

- thermal power of the reactor before shutdown, t_0 - reactor operation time before shutdown (in seconds), $t - t_0$ - time after shutdown (in seconds); the equation is applicable for t in the range: $t_0 + 10$ s $< t < t_0 + 100$ d $\mapsto$ decay heat **D** *Nachwärmeleistung* **F** *puissance de désintégration* **Pl** *moc rozpadu; moc powyłączeniowa* **Sv** *resteffekt*

decay product *rdy* • nuclide that directly or indirectly follows a given radioactive nuclide in a decay chain $\mapsto$ nuclide; radionuclide; decay chain **D** *Folgeprodukt; Zerfallsprodukt; Tochternuklid* **F** *produit de désintégration* **Pl** *produkt rozpadu* **Sv** *sönderfallsprodukt; dotternuklid*

decay series $\mapsto$ decay chain

decay system *rcs* • system for an interim storage of a radioactive material where a dedicated decay of the material can take place $\mapsto$ cooling pond; delay system **D** *Abklingsystem* **F** *système de désactivation* **Pl** *instalacja przechowywania i rozpadu* **Sv** *avklingningssystem*

decay time *rdy* • time during which activity of a radioactive material decreases to the intended level $\mapsto$ activity; radioactive material **D** *Abklingzeit* **F** *temps de désactivation* **Pl** *okres rozpadu* **Sv** *avklingningstid*

decladding $\mapsto$ decanning

decommissioning *roc* • planned shutdown and removal of a nuclear power plant from operation and the work which follows to bring the plant to the final state $\mapsto$ nuclear power plant **D** *Stillegung* **F** *déclassement* **Pl** *wycofywanie z eksploatacji* **Sv** *nedläggning; avveckling*

decontaminate *rdp* • removal of radioactive contamination $\mapsto$ radioactive contamination; decontamination factor **D** *dekontaminieren* **F** *décontaminer* **Pl** *odkażać; dekontaminować* **Sv** *dekontaminera*

decontamination factor *rdp* • ratio

of concentrations of radioactive contamination before and after a cleaning process ↦ decontaminate; radioactive contamination **D** *Dekontaminationsfaktor* **F** *facteur de décontamination* **Pl** *współczynnik dekontaminacji* **Sv** *dekontaminationsfaktor*

defense in depth *rs* • safety principle for construction and operation of nuclear installations that provides guidelines for actions on three levels: accident prevention, accident management and consequence mitigation ↦ nuclear installation; accident management; accident prevention **D** *tiefgestaffeltes Sicherheitsprinzip* **F** *défense en profondeur* **Pl** *obrona w głąb* **Sv** *flernivåskydd; djupförsvar*

deficit ↦*negative reactivity*

degraded core *rs* • deformed core to such an extent that its normal cooling is not possible; the LWR core degradation is usually caused by overheating and can include such stages as: (i) ballooning and rupture of the cladding, (ii) liquefaction and relocation of control and structural materials, (iii) liquefaction and relocation of zircaloy cladding, (iv) liquefaction and slumping the fuel, (v) relocation of molten pool materials into the lower plenum ↦ core overheating **D** *beschädigter Kern* **F** *coeur dégradé* **Pl** *uszkodzony rdzeń* **Sv** *skadad härd*

degree Celsius *th* • (denoted °C:) derived unit of temperature used by the International System of Units SI, exactly equal to one kelvin ↦ Celsius (temperature) scale; kelvin **D** *Grad Celsius* **F** *degré Celsius* **Pl** *stopień Celsjusza* **Sv** *grad Celsius*

degree Fahrenheit *th* • (denoted °F:) unit of temperature in the Fahrenheit scale, where $1 \text{ °F} = \frac{9}{5} \text{ K}$ ↦ Fahrenheit (temperature) scale; Celsius (temperature) scale; kelvin **D** *Grad Fahren-*

heit **F** *degré Fahrenheit* **Pl** *stopień Fahrenheita* **Sv** *grad Fahrenheit*

degree of enrichment *nf* • enrichment factor minus one ↦ enrichment factor **D** *Anreicherungsgrad* **F** *degré d'enrichissement* **Pl** *stopień wzbogacenia* **Sv** *anrikningsgrad*

delayed critical *rph* • such reactor state in which a self-sustained chain reaction is only possible when delayed neutrons are present and available to cause fissions; a formulation used to underline that a self-sustained chain reaction is not possible relying on prompt neutrons only ↦ prompt critical; prompt neutron; delayed neutron **D** *verzögert-kritisch* **F** *critique différé* **Pl** *krytyczny na neutronach opóźnionych* **Sv** *fördröjt kritisk*

delayed neutron *rph* • neutron that appears with an appreciable time delay after the fission event and originates from the subsequent decay of a radioactive fission product ↦ delayed neutron precursor; prompt neutron; fission product **D** *verzögertes Neutron* **F** *neutron retardé* **Pl** *neutron opóźniony* **Sv** *fördröjd neutron*

delayed neutron fraction *rd* • (denoted β:) ratio of the average number of delayed neutrons per fission $\overline{\nu}_d$ to the total (that is, both prompt and delayed) number of neutrons per fission $\overline{\nu}$: $\beta = \overline{\nu}_d / \overline{\nu}$ ↦ delayed neutron; prompt neutron fraction; effective delayed neutron fraction **D** *Anteil der verzögerten Neutronen* **F** *fraction de neutrons retardés* **Pl** *udział neutronów opóźnionych* **Sv** *bråkdel fördröjda neutroner; fördröjd neutronandel*

delayed neutron precursor *rph* • fission fragment whose beta-decay yields a daughter nucleus which subsequently decays via neutron emission, e.g., ^{88}Br, ^{87}Br and ^{137}I; a large number (at least 45) of different *d.n.p.* isotopes will be produced in fission chain reactions, but it has become customary in reactor analysis to group these precursors into six classes characterized by approximate half-lives of 55, 22, 6, 2, 0.5 and 0.2 s ↦ precursor; nuclide; beta decay **D** *Mutterkern verzögerter Neutronen* **F** *précurseur de neutrons retardés* **Pl** *nuklid macierzysty neutronów opóźnionych* **Sv** *föregångare till fördröjda neutroner; föregångare*

delay system *rs* • system with a task to delay a passage of a radioactive material for a long enough time that the required decay can take place ↦ decay system; radioactive material; decay **D** *Verzögerungssystem* **F** *système de désactivation à retardement; système à retardement* **Pl** *układ opóźniający* **Sv** *fördröjningssystem*

delay tank *rs* • tank for interim collection of a radioactive fluid to allow its radioactive decay, until samples can be taken before release of the fluid ↦ delay system; radioactive material; decay **D** *Verweiltank; Abklingtank* **F** *réservoire de désactivation; réservoire de retenue* **Pl** *zbiornik opóźniający* **Sv** *fördröjningstank*

del square ↦*Laplacian operator*

delta-28 *rph* • (for a certain uranium fuel subject to a given neutron spectrum, denoted δ_{28}:) ratio of the total number of fissions in ^{238}U to the number of fissions caused by the thermal neutrons in ^{235}U; this quantity is often set equal to a ratio of the total number of fissions in ^{238}U to the total number of fissions in ^{235}U; the number 28 refers to the last digits in the atomic number (Z=92) and the mass number (A=238) of ^{238}U ↦ fission **D** *Delta-28* **F** *delta-28* **Pl** *delta-28* **Sv** *delta-28*

demand factor *roc* • ratio of the maximum power to the corresponding installed capacity of a turbo-

generator in a nuclear reactor unit ↦ installed capacity **D** *Belastungsfaktor* **F** *facteur de demande; facteur de charge* **Pl** *współczynnik zapotrzebowania* **Sv** *belastningsfaktor*

denatured uranium *nf* • uranium included in a fuel cycle with thorium and ^{233}U, in which the content of ^{233}U is reduced by addition of ^{238}U ↦ fuel cycle; thorium; depleted material **D** *denaturiertes Uran* **F** *uranium dénaturé* **Pl** *uran denaturowany* **Sv** *denaturerat uran*

departure from nucleate boiling *th* • (abbreviated *DNB*:) type of critical heat flux in which the nucleate boiling heat transfer transits into the film boiling heat transfer; the *d.f.n.b.* is mainly relevant to pressurized water reactors, where subcooled nucleate boiling may occur in the fuel assemblies with high power peaking factors; the ratio of the local heat flux q''_{cr} at which the *d.f.n.b.* occurs to the actual local heat flux q''_{ac} in the same fuel assembly is called the *departure from nucleate boiling ratio* and is abbreviated *DNBR*, thus $DNBR = q''_{cr}/q''_{ac}$ ↦ critical heat flux; subcooled nucleate boiling; film boiling; dryout **D** *kritische Überhitzung* **F** *disparition débullition nucléée; crise débullition* **Pl** *odstąpienie od wrzenia pęcherzykowego* **Sv** *uppnående av kritiskt yteffekt*

departure from nucleate boiling ratio ↦*departure from nucleate boiling*

depleted material *mat* • material in which the content of one or several isotopes in a chemical element has been lowered below the normal, naturally occurring level ↦ depletion; depleted uranium; enrichment **D** *verarmtes Material; abgereichertes Material* **F** *matière appauvrie* **Pl** *materiał zubożony* **Sv** *utarmat material*

depleted uranium *mat* • (abbreviated DU:) uranium tail created as a side product in the process of obtaining enriched uranium, in which atomic content of ^{235}U is below 0.7204% ↦ depletion **D** *abgereichertes Uran* **F** *uranium appauvri* **Pl** *uran zubożony* **Sv** *utarmat uran*

depleted zone *mat* • (in material technology:) accumulation of vacancies in a central region of a displacement spike ↦ displacement spike **D** *abgereicherte Zone* **F** *zone appauvrie* **Pl** *strefa zubożona* **Sv** *utarmnad zon*

depletion *nf* • reduction of the content of one or several specified isotopes in a chemical element ↦ isotope; element **D** *Verarmung; Abreicherung* **F** *appauvrissement* **Pl** *zubożenie* **Sv** *utarmning*

deposition (of waste) ↦*controlled tipping*

design-basis accident *rs* • (abbreviated *DBA*:) hypothetical accident conditions against which a nuclear installation is designed according to established design criteria, and for which the damage to the fuel and the release of radioactive material are kept within authorized limits; various hypothetical accidents can be considered as a design basis for a given installation, such as, e.g., breach of the primary system for emergency cooling and containment, or earthquake damage for buildings ↦ nuclear installation; design-basis earthquake **D** *Auslegungsstörfall* **F** *accident de dimensionnement* **Pl** *maksymalna awaria projektowa* **Sv** *konstruktionsstyrande haveri; dimensionerande haveri*

design-basis earthquake *rs* • (abbreviated *DBE*:) earthquake of the maximum intensity taken explicitly into account in the design of a nuclear facility such that the facility can withstand the earthquake with-

out exceeding the authorized limits, assuming the planned operation of safety systems; the *d.b.e.* corresponds to the most severe earthquake that is likely to be experienced from historical records at the site of the facility; in case of a nuclear power plant, an analysis must show that the reactor can be tripped and the safety systems will properly function if such an earthquake occurs ↦ earthquake; safe shutdown earthquake; accident; design-basis accident **D** *Auslegungserdbeben*[1] **F** *séisme normal admissible* **Pl** *maksymalne projektowe trzęsienie ziemi* **Sv** *konstruktionsstyrande jordskalv*

detection time *sfg* • (in safeguards of nuclear materials:) maximum time which can occur between a loss of nuclear material and a detection of the loss ↦ nuclear material **D** *Meßzeit* **F** *temps de détection* **Pl** *czas wykrycia* **Sv** *upptäcktstid*

detector (of radiation) ↦*radiation detector*

detriment *rdp* • expected value of damage in a population due to radiation resulting from operation of certain installations or devices, such as, e.g., a nuclear power plant ↦ radiation; nuclear power plant; collective dose; dose commitment **D** *Strahlenbelastung* **F** *détriment* **Pl** *szkoda (radiacyjna)* **Sv** *detriment*

deuterated water ↦*deuterium*

deuterium *mat* • (denoted $^{2}_{1}$H or D:) stable isotope of hydrogen, with relative atomic mass $A_r = 2.014101778$, occurring naturally on Earth in a molecule HDO of the *deuterated water*, which is present in ordinary water at an abundance of approximately 1 molecule in 3200 ↦ heavy water; hydrogen; isotope; relative atomic mass **D** *Deu-*

terium **F** *deutérium* **Pl** *deuter; ciężki wodór* **Sv** *deuterium*

deuteron *bph* ● (denoted *d*:) nucleus of deuterium that is a bound state of proton and neutron, with a binding energy of 2.2 MeV; the only nuclide with the mass number $A = 2$ ↦ deuterium **D** *Deuteron* **F** *deutéron* **Pl** *deuteron* **Sv** *deutron*

differential cross section *xr* ● cross section, calculated per angle or energy, which characterizes the probability that a collision reaction changes the original particle energy E and direction angle Ω to a certain new energy from E' to $E'+dE'$ and direction angle Ω' in $d\Omega'$ ↦ cross section; angular cross section **D** *differentieller Wirkungsquerschnitt* **F** *section efficace différentielle* **Pl** *przekrój czynny różniczkowy* **Sv** *differentiellt tvärsnitt*

differential dose albedo *rdp* ● reflection factor of a surface for radiation, valid for reflection in a certain direction and expressed as a ratio of two dose rates ↦ dose rate; albedo **D** *differentielle Dosisalbedo* **F** *albédo de dose différentiel* **Pl** *albedo dawki różniczkowej* **Sv** *differentiell dosalbedo*

differential reactivity *rph* ● (also called *differential worth*:) change of reactivity per control rod displacement ↦ reactivity; integral reactivity **D** *differentielle Reaktivität* **F** *réactivité différentielle* **Pl** *reaktywność różniczkowa* **Sv** *differentiell reaktivitet*

differential worth ↦*differential reactivity*

diffuser *nf* ● (for isotope separation:) component containing the separating elements in a stage in a gas diffusion cascade ↦ isotope separation **D** *Diffusor* **F** *diffuseur* **Pl** *dyfuzor* **Sv** *diffusor*

diffusion area *rph* ● one-sixth of the mean square distance between the point where a particle (e.g., a neutron) of a certain type or class appears and the point where the particle disappears from the type or the class in an infinite homogeneous medium; the *d.a.* is equal to the diffusion length squared ↦ diffusion length; migration area **D** *Diffusionsfläche* **F** *aire de diffusion* **Pl** *powierzchnia dyfuzji* **Sv** *diffusionsarea*

diffusion barrier *nf* ● (for isotope separation:) porous membrane in which small pore sizes hinder the ordinary gas flow but allow diffusion ↦ isotope separation **D** *Diffusionsmembrane; Diffusionsbarriere* **F** *barrière de diffusion* **Pl** *bariera dyfuzyjna* **Sv** *diffusionsbarriär*

diffusion coefficient *rph* ● (also called *neutron diffusion coefficient*, denoted D:) proportionality factor in Fick's approximation used in the neutron diffusion theory; the *d.c.* is defined as $D = (3\Sigma_{tr})^{-1}$, where Σ_{tr} is the transport cross section ↦ Fick's law; neutron diffusion; transport cross section **D** *Diffusionskoeffizient* **F** *coefficient de diffusion* **Pl** *współczynnik dyfuzji* **Sv** *diffusionskoefficient*

diffusion cooling *rph* ● (for neutron diffusion in a finite body:) reduction of mean energy of neutrons caused by leakage of neutrons with higher energy from the body ↦ neutron diffusion **D** *Diffusionskühlung* **F** *refroidissement par diffusion* **Pl** *chłodzenie dyfuzyjne* **Sv** *diffusionskylning*

diffusion equation *rph* ● partial differential equation that describes the diffusion of particles according to the diffusion theory; in its simplest form, referred to as the *one-group diffusion equation*, the *d.e.* is as follows $\frac{1}{v}\frac{\partial\phi}{\partial t} = -\nabla\cdot D\nabla\phi + \Sigma_a\phi = S$, where v - particle speed, ϕ - one-group flux density, D - one-group diffusion coefficient, Σ_a - macroscopic absorption cross section, S - particle source and t - time $\mapsto$ diffusion theory D *Diffusionsgleichung* F *équation de la diffusion* Pl *równanie dyfuzji* Sv *diffusionsekvation*

diffusion length *rph* • (denoted L:) $L \equiv \sqrt{D/\Sigma_a}$, where D - diffusion coefficient, Σ_a - absorption cross section; for diffusion of monoenergetic neutrons from a point source it can be shown that $L^2 = \frac{1}{6}\overline{r^2}$, where $\overline{r^2}$ is the mean square distance from the source to the point at which the neutron is absorbed $\mapsto$ diffusion area D *Diffusionslänge* F *longueur de diffusion* Pl *długość dyfuzji (neutronów monoenergetycznych)* Sv *diffusionslängd*

diffusion theory *rph* • approximate theory of diffusion of particles, e.g., neutrons, based on the assumption that the current density is proportional to the gradient of the particle flux density in a homogeneous medium; the conservation of particles is expressed by the diffusion equation with the scalar particle flux density as the dependent variable $\mapsto$ diffusion equation; particle flux density D *Diffusionstheorie* F *théorie de la diffusion* Pl *teoria dyfuzji* Sv *diffusionsteori*

direct-cycle boiling reactor *rty* • boiling water reactor in which the vapour generated in the reactor core is flowing directly to the turbine $\mapsto$ boiling water reactor; direct-cycle reactor D *Direktkreissiederaktor* F *réacteur bouillant à cycle direct* Pl *reaktor wrzący z obiegiem bezpośrednim* Sv *direktcykelkokare*

direct-cycle reactor *rty* • reactor in which the primary circuit coolant is directly used to generate the desired form of energy without applying a secondary circuit coolant $\mapsto$ primary-coolant circuit; secondary-coolant circuit; coolant D *Direktkreisraktor* F *réacteur à cycle direct* Pl *reaktor z obiegiem bezpośred-*

nim Sv *direktcykelrektor; reaktor med direkt cykel*

direct disposal *wst* • ultimate waste disposal of spent fuel without fuel reprocessing $\mapsto$ ultimate waste disposal; spent fuel; fuel reprocessing D *direkte Endlagerung* F *évacuation directe* Pl *składowanie ostateczne bezpośrednie* Sv *direktförvaring*

directional dose equivalent *rdp* • (denoted H*(0.07), measured in sieverts:) dose equivalent at 0.07 mm depth in a sphere with 30 cm diameter made of tissue-like material (with density 1000 kg/m^3) in a specified direction; *d.d.e.* is measurable and it represents the most reliable approximation of the dose equivalent for the skin at 0.07 mm depth $\mapsto$ dose equivalent; sievert D - F *équivalent de dose directionelle* Pl *kierunkowy równoważnik dawki pochłoniętej* Sv *riktningsdosekvivalent*

directly ionizing particle *rd* • charged particle with high-enough kinetic energy to cause ionization during a collision, such as, e.g., electron, proton or alpha particle $\mapsto$ ionize; indirectly ionizing particle; electron; proton; alpha particle D *direkt ionisierendes Teilchen* F *particule directment ionisante* Pl *cząstka bezpośrednio jonizująca* Sv *direkt joniserande partikel*

direct maintenance *roc* • inspection and maintenance through direct, manual actions $\mapsto$ remote maintenance D *direkte Handhabung* F *maintenance directe* Pl *konserwacja i nadzór bezpośredni* Sv *närtillsyn*

disadvantage factor *rph* • (for material in a reactor cell:) ratio of the mean value of neutron flux density in the material and in the fuel; in particular the *thermal disadvantage factor* is defined as the ratio of the average thermal neutron flux in the moderator to that in the fuel; the ABH (Amouyal, Benoist and Horowitz) method can be used to determine the *d.f.* ↦ neutron flux density *D* *Absenkungsfaktor; Absenkungsverhältnis* *F* *facteur de désavantage* *Pl* *współczynnik niekorzyści* *Sv* *depressionsfaktor*

discrete ordinates method *rph* • numerical method to solve the neutron transport equation by division of the spatial angle into *N* segments and approximation of the differential particle flux density with a linear combination of its values in certain directions ↦ transport equation; particle flux density; finite element method *D* *Carlsonsche S_N-Methode* *F* *méthode de Carlson (S_N)* *Pl* *metoda dyskretnych współrzędnych S_N* *Sv* *S_N-metod*

dishing *nf* • cavity in one or both ends of fuel pellets to reduce length expansion of the pellet's pillar ↦ fuel pellet; axial fuel gap *D* *Wölben* *F* *concavité; évidement sphérique* *Pl* *miseczkowanie* *Sv* *skålning*

disintegration constant ↦*decay constant*

disintegration rate *rdy* • number of radioactive decays per unit time ↦ radioactive decay; activity *D* *Zerfallsrate* *F* *taux de désintégration* *Pl* *szybkość rozpadu* *Sv* *sönderfallsrat*

dispersion *wst* • dilution of a waste substance due to transport, diffusion and mixing ↦ nuclear waste *D* *Aufstreuung* *F* *dispersion* *Pl* *rozproszenie* *Sv* *spridning*[1]

dispersion fuel *nf* • nuclear fuel in a form of particles dispersed in another material ↦ nuclear fuel *D* *dispergierter Brennstoff* *F* *combustible en dispersion* *Pl* *paliwo rozproszone* *Sv* *dispersionsbränsle*

dispersion reactor *rty* • nuclear reactor using the dispersion fuel ↦ dispersion fuel *D* *Dispersionsreaktor* *F* *réac-*

teur à combustible en dispersion *Pl* *reaktor z paliwem rozproszonym* *Sv* *dispersionsreaktor*

displacement spike *mat* • radiation damage zone (spike) consisting of a number of atoms that are temporarily or permanently expelled from their normal position by a primary knocked-on atom or another directly ionizing particle; the expelled atoms leave vacancies at the center of the *d.s.*, which together create a depleted zone ↦ spike; depleted zone *D* *Verschiebungsstörzone* *F* *zone de déplacements (dégâts par rayonnements)* *Pl* *szczyt przemieszczeń* *Sv* *utstötningszon*

dissolver *nf* • (in fuel reprocessing:) vessel for selective solution of the fuel material through leaching with an acid ↦ fuel reprocessing *D* *Auflöser* *F* *dissolveur* *Pl* *zbiornik do rozpuszczania* *Sv* *upplösare*

Dittus-Boelter correlation *th* • experimental correlation used to determine the Nusselt number as a function of the Reynolds and Prandtl numbers; the *D.-B.c.* is valid for turbulent heat transfer to non-metallic liquids and gases in round tubes; the correlation is given as: $Nu = 0.023Re^{0.8}Pr^n$ where $n = 0.4$ when the fluid is heated and $n = 0.3$ when the fluid is cooled; Nu is the Nusselt number, Re is the Reynolds number and Pr is the Prandtl number; the *D.-B.c.* is valid when $L/D > 60, Re > 10^4$ and $0.7 < Pr < 100$, where L is the distance from the tube entrance and D is the tube diameter ↦ Colburn correlation; Nusselt number; Prandtl number; Reynolds number *D* *Dittus-Boelter Gleichung* *F* *corrélation de Dittus-Boelter* *Pl* *korelacja Dittusa-Boeltera* *Sv* *Dittus-Boelters samband*

diversification *rs* • construction

46

principle to increase reliability of a safety function, which can be achieved by at least two systems with different principles of operation ↦ safety function; defense in depth *D* *Diversifikation* *F* *diversification* *Pl* *zróżnicowanie* *Sv* *diversifiering*

DNB ↦*departure from nucleate boiling*

DNB correlation *th* • experimentally derived formula that expresses the critical heat flux for film boiling in terms of various characteristic parameters for the coolant channel, such as, e.g., $q''_{cr} = f(G, p, x_e)$, where q''_{cr} is the critical heat flux, G is the mass flux, p is the pressure and x_e is the equilibrium thermodynamic quality in the coolant channel ↦ departure from nucleate boiling; critical heat flux; film boiling; coolant channel *D* *Beziehung für kritische Wärmestromdichte* *F* *corrélation de caléfaction* *Pl* *korelacja odstąpienia od wrzenia pęcherzykowego* *Sv* *filmkokningskorrelation*

DNBR ↦*DNB ratio*

DNB ratio *th* • (for a specific point on the cladding, abbreviated *DNBR:*) ratio of the critical heat flux for film boiling and the actual heat flux ↦ departure from nucleate boiling; critical heat flux; film boiling; heat flux *D* *Durchbrennmarge* *F* *marge de caléfaction* *Pl* *współczynnik odstąpienia od wrzenia pęcherzykowego* *Sv* *filmkokningskvot; filmkokningsmarginal*

dollar *rph* • reactivity unit that is equivalent to the amount of reactivity necessary to make a reactor prompt critical: $\rho = \beta$, where β is the delayed neutron fraction; since the value of β varies from fuel to fuel, the *d.* is not an absolute unit ↦ reactivity; cent; prompt neutron; delayed neutron fraction *D* *Dollar* *F* *dollar* *Pl* *dolar* *Sv* *dollar*

Doppler broadening *xr* • (in reactor physics:) observed increase of width of a cross section resonance caused by thermal movements of target particles ↦ cross section *D* *Doppler-Verbreiterung* *F* *élargissement Doppler* *Pl* *rozszerzenie Dopplera* *Sv* *dopplerbreddning*

Doppler coefficient *rph* • fraction of the fuel temperature coefficient of reactivity which results from the Doppler broadening ↦ temperature coefficient of reactivity; Doppler broadening *D* *Doppler-Koeffizient* *F* *coefficient Doppler* *Pl* *współczynnik Dopplera* *Sv* *dopplerkoefficient*

Doppler effect *rph* • (in reactor physics:) situation when even modest speeds of nuclear thermal motion can significantly affect the energy dependence of the neutron cross section in the vicinity of the resonance; this phenomenon is akin to a frequency shift that accompanies variations in relative motions between source and receiver in sound propagation ↦ Doppler broadening *D* *Doppler-Effekt* *F* *effet Doppler* *Pl* *zjawisko Dopplera* *Sv* *dopplereffekt*

dose *rdp* • expression referring to an absorbed energy from a radiation field ↦ dose equivalent; dose rate; dose commitment *D* *Dosis* *F* *dose* *Pl* *dawka* *Sv* *dos; stråldos*

dose albedo *rdp* • reflection factor of a surface for radiation expressed as a ratio of two dose rates ↦ radiation; albedo; dose; dose rate *D* *Dosisalbedo* *F* *albédo de dose* *Pl* *albedo dawki* *Sv* *dosalbedo*

dose commitment *rdp* • time integral over all coming years of the effective dose rate equivalent to individuals in a certain population after a certain measure which entails risks for irradiation, such as, e.g., emissions of radioactive materials; the population can be the whole Earth's population or a limited group and it doesn't have to consist of the same individuals over years; the unit of *d.c.* is the sievert ↦ irradiation; radioactive material; collective dose; sievert **D** *resultierende Dosis* **F** *engagement de dose absorbé* **Pl** *dawka wynikowa* **Sv** *dosinteckning*

dose equivalent *rdp* ● (only within radiation protection:) product of an absorbed dose, a quality factor, and other factors which are necessary to estimate the radiation's effect on exposed individuals; the unit of *d.e.* is the sievert ↦ absorbed dose; quality factor; sievert **D** *Äquivalentdosis* **F** *équivalent de dose* **Pl** *równoważnik dawki* **Sv** *dosekvivalent*

dosemeter ↦*dosimeter*

dose rate *rdp* ● absorbed dose per unit time ↦ absorbed dose **D** *Dosisleistung* **F** *débit de dose* **Pl** *moc dawki* **Sv** *dosrat*

dose-rate meter *rdp* ● instrument to measure the dose rate ↦ dose rate **D** *Dosisleistungsmesser* **F** *débitmètre de dose* **Pl** *miernik mocy dawki* **Sv** *dosratsmätare*

dosimeter *rdp* ● meter of the absorbed dose, exposure or equivalent radiation quantities, equipped with a sensor and sometimes even with a reading instrument ↦ absorbed dose; exposure **D** *Dosimeter; Dosismesser* **F** *dosimètre* **Pl** *dozymetr; dawkomierz* **Sv** *dosimeter; dosmätare*

dosimetry *rdp* ● measuring of various physical parameters that characterise the influence of ionizing radiation on matter, such as, e.g., the exposure or the absorbed dose ↦ ionizing radiation; exposure; absorbed dose **D** *Dosimetrie* **F** *dosimétrie* **Pl** *dozymetria* **Sv** *dosimetri*

double spherical harmonics method *rph* ● approximate method to solve the transport equation based on double series expansions of the differential angular distribution of the particle flux density: one in the for-

ward direction and one in the backward direction; the method is particularly useful at surfaces between two materials with significantly different neutron properties ↦ spherical harmonics method; particle flux density **D** *Methode der Kugelfunktionsentwicklung im Halbraum; Doppel-P_N-Methode* **F** *méthode des doubles harmoniques sphériques; méthode d'Yvon* **Pl** *metoda funkcji podwójnych sferycznych harmonicznych; metoda Yvona* **Sv** *dubbel klotfunktionsmetod; Yvons metod*

doubling time[1] *roc* ● time interval during which a physical quantity such as, e.g., power of a nuclear reactor, is doubled; *d.t.* is usually used in exponential processes ↦ doubling time[2]; doubling time[3] **D** *Verdopplungszeit[1]* **F** *temps de doublement[1]* **Pl** *czas podwojenia[1]* **Sv** *fördubblingstid[1]*

doubling time[2] *rph* ● (within breeding reactor technology:) time interval during which the initial amount of fissile nuclides in a certain fuel load of a breeder reactor is doubled through breeding at normal reactor operation; for a breeding reactor with an initial fuel inventory M_0 [kg], breeding gain G, thermal rated power P [MWth], fraction of time at rated power f and mean capture-to-fission ratio $\bar{\alpha}$, *d.t.* can be estimated as,

$$d.t. \cong \frac{2.73 M_0}{G \cdot P \cdot f \cdot (1 + \bar{\alpha})}$$

where *d.t.* is in years ↦ doubling time[1]; doubling time[3]; breeding **D** *Verdopplungszeit[2]* **F** *temps de doublement[2]* **Pl** *czas podwojenia[2]* **Sv** *fördubblingstid[2]*

doubling time[3] *rph* ● (within breeding reactor technology:) corresponding time interval as doubling time[2] for the fissile nuclides in a fuel cycle for a breeder reactor, possibly along with several other nuclear reactors ↦

doubling time[1]; doubling time[2]; breeding **D** *Verdopplungszeit*[3] **F** *temps de doublement*[3] **Pl** *czas podwojenia*[3] **Sv** *fördubblingstid*[3]

downcomer *rcs* • space in a nuclear reactor vessel in which the circulating coolant flows down and which in light-water reactors consists of a gap between the reactor pressure vessel and the moderator tank ↦ coolant; reactor pressure vessel; moderator tank **D** *Ringraum zur Kühlmittelniederführung* **F** *espace annulaire (pour les réacteurs à eau ordinaire)* **Pl** *szczelina opadowa* **Sv** *fallspalt*

driver fuel *nf* • nuclear fuel in the driver zone ↦ driver zone **D** *Treiberbrennstoff* **F** *combustible nourriciér* **Pl** *paliwo zasilające* **Sv** *drivbränsle*

driver zone *rph* • (in a multi-zone core:) zone in a nuclear reactor core, which mainly maintains the chain reaction ↦ driver fuel **D** *Treiberzone* **F** *zone nourricière* **Pl** *strefa zasilająca* **Sv** *drivzon*

drum *wst* • (for waste management:) container which has roughly the same shape and volume as an oil drum and which can be sealed with a lid; the *d.* can be enclosed in concrete for radiation shielding ↦ waste management; radiation shielding **D** *Faß* **F** *fût*[2] **Pl** *walczak* **Sv** *fat*

dryout *th* • type of critical heat flux (CHF), which occurs in channels with high content of the vapour phase; *d.* occurs when a liquid film evaporates and disappears from a heated wall; in heat-flux-controlled heaters, the wall temperature significantly increases due to a sudden deterioration of the heat transfer rate ↦ dryout correlation; departure from nucleate boiling; critical heat flux **D** *Übergang zum Filmsieden* **F** *assèchement* **Pl** *wyschnięcie ścianki* **Sv** *torrkokning*

dryout correlation *th* • experimentally derived relationship between the critical heat flux causing dryout and different parameters characterizing the coolant channel, usually given as $x_{cr} = f(G, p, D_h, L_B, F_{int}, ...)$, where x_{cr} - critical quality, G - mass flux of coolant, p - system pressure, D_h - channel hydraulic diameter, L_B - boiling length, F_{int} - radial peaking factor ↦ dryout; critical heat flux; coolant channel; hydraulic diameter; radial peaking factor; dryout margin **D** *Beziehung für kritische Heizflächenbelastung* **F** *corrélation de dénoyage* **Pl** *korelacja wyschnięcia ścianki* **Sv** *torrkokningskorrelation*

dryout margin *th* • (for a coolant channel:) difference between the critical coolant channel power at which the dryout occurs and the actual coolant channel power ↦ coolant channel; critical power ratio; dryout; dryout correlation **D** *Abstand zur kritischen Heizflächenbelastung* **F** *facteur de sécurité de dénoyage* **Pl** *zapas mocy do wyschnięcia ścianki* **Sv** *torrkokningsmarginal*

drywell *rcs* • space inside the containment of a boiling water reactor used for expansion of vapour released during a loss-of-coolant accident in the primary-coolant circuit ↦ boiling water reactor; containment; loss-of-coolant accident; primary-coolant circuit; pedestal **D** *Expansionsraum* **F** *volume d'expansion* **Pl** *komora rozprężeniowa* **Sv** *primärutrymme*

dual-cycle reactor *rty* • reactor in which a useful power is obtained through use of heat from both the primary and the secondary-coolant circuit ↦ primary-coolant circuit; secondary-coolant circuit **D** *Zweikreisreraktor* **F** *réacteur à double cycle* **Pl** *reaktor dwuobiegowy* **Sv** *tvåkretsreaktor*

dual-pressure cycle *th* • steam cycle in which the steam is generated and used at two different pressures

↦ Rankine cycle; Brayton cycle **D** *Zwei-druckkreis* **F** *cycle à double pression* **Pl** *obieg parowy dwuciśnieniowy* **Sv** *tvåtryckscykel*

dual temperature exchange separation process *nf* • isotope separation process based on the fact that for two interacting chemical compounds of the same chemical element, the heavier isotope usually concentrates in the less-volatile compound, which is kept at a lower temperature; the *d.t.e.s.p.* is used in the production of heavy water ↦ isotope separation; heavy water **D** *Zweitemperatur-Austausch-Trennverfahren* **F** *procédé bitherme de séparation* **Pl** *dwutemperaturowy proces separacji (izotopów)* **Sv** *tvåtemperaturprocess*

dummy assembly *nf* • part of a nuclear reactor core that is similar to the nuclear fuel assembly, but does not contain any nuclear fuel; the *d.a.* is used for testing purposes ↦ dummy element; fuel assembly; reactor core **D** *Blin-delementbündel* **F** *assemblage postiche; faux assemblage* **Pl** *atrapa kasety* **Sv** *blindpatron*

dummy element *nf* • part of a nuclear reactor core that is similar to the nuclear fuel element, but does not contain any nuclear fuel; the *d.e.* is used for testing purposes ↦ dummy assembly; fuel element **D** *Blindelement* **F** *élément postiche; faux élément* **Pl** *atrapa elementu* **Sv** *blindelement*

duty engineer *roc* • person who is ready to take over the responsibility for operation of an installation in case of an emergency situation ↦ reference level for emergency action **D** *Ingenieur mit Bereitschaftsdienst* **F** *ingénieur d'astreinte* **Pl** *dyżurny inżynier* **Sv** *vakthavande ingenjör*

dyne *bph* • old unit of force equal to 10^{-5} N, defined as the force required to accelerate a mass of one gram at a rate of one centimetre per second squared ↦ newton **D** *Dyn* **F** *dyne* **Pl** *dyna* **Sv** *dyn*

Ee

EAL $\mapsto$*reference level for emergency action*

earthquake *rs* • one of naturally occurring phenomena considered in safety analyses of nuclear installations; the *e.* is a significant shaking of the Earth's surface, resulting from the sudden release of energy in its crust; in many countries the reactor plant design is considered to be satisfactory if it is designed to withstand a maximum horizontal acceleration of 0.3 g $\mapsto$ nuclear installation; safe shutdown earthquake; design-basis earthquake; tsunami; log-normal distribution; Fukushima Daiichi nuclear accident **D** *Erdbeben* **F** *séisme* **Pl** *trzęsienie ziemi* **Sv** *jordbävning*

ECCS $\mapsto$*emergency core cooling system*

effective delayed neutron fraction *rph* • ratio of the mean number of fissions caused by delayed neutrons to the total mean number of fissions caused by both delayed and prompt neutrons $\mapsto$ fission; delayed neutron fraction; delayed neutron; prompt neutron **D** *effektiver Anteil der verzögerten Neutronen* **F** *fraction efficace de neutrons retardés* **Pl** *efektywna wydajność neutronów opóźnionych* **Sv** *effektiv bråkdel fördröjda neutroner; effektiv fördröjd neutronandel*

effective dose equivalent *rdp* • sum of a product of the dose equivalent and the weighting factor for various organs in the body; the *e.d.e.* represents the best measure of the cancer risk due to irradiation; the weighting factors are provided by the International Commission on Radiological Protection $\mapsto$ dose equivalent; International Commission on Radiological Protection **D** *effektive Äquivalentdosis* **F** *équivalent de dose effectif* **Pl** *równoważnik dawki skutecznej* **Sv** *effektiv dosekvivalent*

effective full-power hours *roc* • (for a power reactor, abbreviated *EFPH*:) ratio of the generated energy during a certain time period and the rated power of the reactor; the term is applicable to the electrical gross or net power, or to the thermal power $\mapsto$ power reactor; electrical power; thermal power **D** *effektive Vollastzeit* **F** *heures effectives à pleine puissance* **Pl** *równoważny okres pełnej mocy; efektywny czas pełnej mocy* **Sv** *ekvivalent fulleffekttid*

effective half-life *rdp* • time during which the amount of a radioactive nuclide in a system reduces to one-half of the initial value, both due to radioactive decay and other processes, e.g., the biological decay $\mapsto$ radioactive half-life; biological half-life **D** *effektive Halbwertzeit* **F** *période effective* **Pl** *równoważny okres półrozpadu* **Sv** *effektiv halveringstid*

effective kilogram *sfg* • quantity used in safeguards of nuclear materials to represent the strategic value of the material $\mapsto$ safeguards; nuclear material **D** *effektives Kilogramm* **F** *kilogramme* *effectif* **Pl** *kilogram równoważny* **Sv** *effektiv massa*

effective multiplication constant
↦*effective multiplication factor*

effective multiplication factor *rph* • (denoted k_{eff}:) multiplication factor for an infinite system ↦ multiplication factor **D** *effektiver Multiplikationsfaktor; effektiver Vermehrungsfaktor* **F** *facteur de multiplication effectif* **Pl** *efektywny współczynnik mnożenia neutronów* **Sv** *effektiv multiplikationskonstant; kriticitetsfaktor*

effective resonance integral *rph* • resonance integral in which the cross section has been replaced with the effective cross section that gives the correct reaction rate when the neutron flux density is not inversely proportional to the neutron energy ↦ resonance integral; cross section; reaction rate; neutron flux density **D** *effektives Resonanzintegral* **F** *intégrale effective de résonance* **Pl** *efektywna całka rezonansowa* **Sv** *effektiv resonansintegral*

effective thermal cross section *xr* • (denoted σ_{eff}:) average cross section calculated for a certain reaction in such a way that, when multiplied with the conventional flux density, it gives the correct reaction rate ↦ thermal cross section; conventional flux density; reaction rate; Westcott model **D** *effektiver thermischer Wirkungsquerschnitt* **F** *section efficace thermique effective* **Pl** *efektywny termiczny przekrój czynny* **Sv** *effektivt termiskt tvärsnitt; Westcott-tvärsnitt*

efficiency coefficient *th* • (for energy transformation systems, denoted η:) ratio of the useful energy yielded by a device or system to the total energy provided to the same device or system ↦ internal consumption; net power **D** *Wirkungsgrad* **F** *coefficient de rendement* **Pl** *współczynnik sprawności* **Sv** *verkningsgrad*

EFPH ↦*effective full-power hours*

eigenvalue *mth* • scalar λ that satisfies equation $T(\vec{v}) = \lambda\vec{v}$, where $\vec{v}$ is a non-zero vector (referred to as the *eigenvector*) and T is a linear transformation ↦ multiplication factor **D** *Eigenwert* **F** *valeur propre* **Pl** *wartość własna* **Sv** *egenvärde*

eigenvector ↦*eigenvalue*

elastic scattering *nap* • scattering of a particle during which the total kinetic energy of the particle is preserved ↦ scattering; inelastic scattering **D** *elastische Streuung* **F** *diffusion élastique* **Pl** *rozpraszanie sprężyste* **Sv** *elastisk spridning*

electrical power *roc* • gross or net power of a reactor unit that is distributed to an electrical power grid ↦ reactor unit; gross power; net power; thermal power **D** *elektrische Leistung* **F** *puissance électrique* **Pl** *moc elektryczna* **Sv** *elektrisk effekt*

electromagnetic separation process *nf* • isotope separation process resulting from the dependence of an ion deflection in a magnetic field on the ratio of the ion's mass to its electrical charge ↦ isotope separation **D** *elektromagnetisches Trennverfahren* **F** *séparation électromagnétique* **Pl** *separacja elektromagnetyczna* **Sv** *elektromagnetisk separationsprocess*

electron *bph* • (denoted e^- or β^-:) stable subatomic particle with the rest mass $m_e = 9.109\,383\,56(11) \times 10^{-31}$ kg $\cong 0.5110$ MeV/c$^2 \cong 5.486 \times 10^{-4}$ u and with a negative elementary charge e ↦ atom; atomic mass unit; elementary charge **D** *Elektron* **F** *électron* **Pl** *elektron* **Sv** *elektron*

electron emission detector ↦*collectron*

electron volt *bph* • (denoted eV:) energy unit, dedicated to nuclear physics and related fields, equal to the kinetic energy gained by an electron while passing in a vacuum between two points, when the potential difference between the points is equal to 1 V; 1 eV = 1.602 176 487(40)×10^{-19} J; derived units of eV include: 1 keV = 10^3 eV, 1 MeV = 10^6 eV and 1 GeV = 10^9 eV ↦ electron **D** *Elektronenvolt* **F** *électronvolt* **Pl** *elektronowolt* **Sv** *elektronvolt*

element *bph* • substance which contains atoms having the same atomic number Z and the same chemical properties; there are 90 *e.* existing in nature, with Z from 1 to 92; elements with $Z = 43$ and 61, and all with $Z > 92$ are synthetic; majority of the naturally occurring elements are composed of two or more isotopes; the isotopic abundance γ_i of i-th isotope in a given *e.* is the fraction of atoms in the *e.* that are of i-th type; the relative atomic mass of the *e.* is obtained as a weighted average of the relative atomic masses of the isotopes as follows,

$$A_r = \sum_i \frac{\gamma_i(\%)}{100} A_{r,i}$$

here $A_{r,i}$ is the relative atomic mass of i-th isotope and A_r is the relative atomic mass of the *e.* ↦ atomic number; relative atomic mass; isotope; isotopic abundance; atom **D** *Element* **F** *élément* **Pl** *pierwiastek* **Sv** *grundämne; element*

elementary charge *bph* • (usually denoted *e:*) one of the exact fundamental physical constants equal to $e = 1.602\ 176\ 634 \cdot 10^{-19}$ C (exact), used in a definition of the ampere A = C/s; electric charge carried by a single proton (if positive), or a single electron (if negative) ↦ electron; proton; electron volt **D** *Elementarladung* **F**

charge *élémentaire* **Pl** *ładunek elementarny* **Sv** *elementarladdning*

elementary particle *bph* • traditional name of sub-nuclear particles; according to the current Standard Model, there are 17 elementary particles, including 12 fermions (6 leptons and 6 quarks), 4 force carriers and the Higgs boson ↦ atom; nucleus; neutron **D** *Elementarteilchen* **F** *particule élémentaire* **Pl** *cząstka elementarna* **Sv** *elementarpartikel*

element for automatic control ↦*member for automatic control*

emergency action level ↦*reference level for emergency action*

emergency class ↦*reference level for emergency action*

emergency containment cooling *rcs* • spray system for cooling of the containment in case of an abnormal heat release due to, e.g., release of the vapour and hot water from a breach in the primary coolant circuit ↦ containment; primary-coolant circuit **D** *Notkühlsystem für den Sicherheitsbehälter* **F** *système de refroidissement de secours de l'enceinte confinement* **Pl** *awaryjne chłodzenie obudowy bezpieczeństwa* **Sv** *nödkylsystem för reaktorinneslutning; sprinklersystem för reaktorinneslutning*

emergency core cooling *th* • evacuation of the decay heat from a reactor core after the loss-of-coolant accident; the *e.c.c.* can be realised either through spray cooling or through reflooding ↦ loss-of-coolant accident; spray cooling; reflooding **D** *Notkühlung des Reaktorkerns* **F** *refroidissement d'urgence du cœur* **Pl** *awaryjne chłodzenie rdzenia* **Sv** *härdnödkylning*

emergency core cooling system *rcs* • (abbreviated *ECCS:*) system for cooling of a reactor core after the loss of primary coolant; in a light-water reactor the *e.c.c.s.* usually consists of the high-pressure system and the low-pressure system; in a pressurized water reactor plant the *e.c.c.s.* contains in addition the accumulator system ↦ primary coolant; light-water reactor; accumulator system **D** *Notkühlsystem für den Reaktorkern* **F** *système de refroidissement de secours du coeur* **Pl** *układ awaryjnego chłodzenia rdzenia* **Sv** *nödkylsystem för härd; härdnödkylsystem*

emergency planning zone *rdp* ● (abbreviated *EPZ*:) area surrounding a nuclear installation where a preparedness plan exists; typically *e.p.z.* stretches to 12–15 km from the installation; two types of emergency zones are defined: (i) precautionary action zone, in which protective actions are to be taken before or shortly after a release of radioactive material; (ii) urgent protective action planning zone, in which urgent protective actions are taken based on environmental monitoring ↦ nuclear installation; emergency preparedness; exclusion area **D** *Zone für Notfallplanung* **F** *zone de planification d'urgence* **Pl** *strefa planowania awaryjnego* **Sv** *inre beredskapszon*

emergency preparedness *rs* ● planned and exercised measures inside and outside of a nuclear installation to limit consequences of a nuclear accident ↦ nuclear installation; accident management; accident prevention; nuclear accident **D** *Notfallplanung* **F** *plans d'intervention en cas d'urgence* **Pl** *gotowość na wypadek awarii (jądrowej)* **Sv** *haveriberedskap*

emergency shutdown limit *rs* ● limit value of an operational parameter which causes a reactor trip when exceeded ↦ reactor trip **D** *Notabschaltungsgrenze* **F** *limite d'arrêt d'urgence* **Pl**

próg awaryjnego wyłączenia reaktora **Sv** *snabbstoppsgräns*

emission rate *rd* ● number of particles that leave a radiation source per unit time; the *e.r.* can refer to all particles or to particles of a certain kind and with a certain energy ↦ radiation source **D** *Emissionsrate; Quellstärke* **F** *débit* **Pl** *częstość emisji* **Sv** *källstyrka; emissionsrat*

encapsulation *wst* ● procedure which is used to provide, e.g. radioactive wastes, with a closed casing ↦ radioactive waste; canning **D** *Einkapseln; Einkapselung* **F** *gainage^2*; encapsulation **Pl** *kapsułkowanie* **Sv** *inkapsling*

end of cycle *roc* ● (abbreviated *EOC*:) time shortly before refuelling, used as a reference time to determine the fuel condition ↦ refuelling **D** *Zyklusende* **F** *fin du cycle* **Pl** *koniec cyklu paliwowego* **Sv** *slutet av driftperiod*

energy *th* ● property of a body that can be transferred either by heat or work; various energy forms are distinguished, such as: kinetic energy, potential energy, thermal energy, etc.; depending on its origin, the energy can be termed as fossil energy, hydro energy, solar energy, geothermal energy, nuclear energy, etc.; the SI unit of energy is the joule ↦ joule **D** *Energie* **F** *énergie* **Pl** *energia* **Sv** *energi*

energy availability factor *roc* ● (for a given reactor unit:) ratio of the gross electricity production during a certain time to a product of the total calendar time and the nominal gross power, expressed in percent ↦ gross power; reactor unit **D** *Arbeitsverfügbarkeit* **F** *taux de disponibilité en énergie* **Pl** *wskaźnik dyspozycyjności eneregetycznej* **Sv** *energitillgänglihetsfaktor*

energy imparted to matter *rdp* ● sum of the energies (excluding rest-mass energies) of all charged and un-charged particles entering a certain volume with mass m minus the sum of the energies (excluding the rest-mass energies) of all charged and un-charged ionizing particles leaving the volume, further corrected by subtract-ing the energy equivalent of any in-crease in rest-mass energy in the vol-ume; *e.i.t.m.* is involved in the ion-ization and excitation of atoms and molecules and is eventually degraded almost entirely into the thermal en-ergy ↦ absorbed dose; kerma; nuclear re-action **D** *auf das Material übertragene En-ergie* **F** *énergie communiquée à la matière* **Pl** *zaabsorbowana energia promieniowania* **Sv** *absorberad strålningsenergi; absorberad en-ergi*

engineered safety *rs* • feature of a component or a system which pre-vents a certain potentially dangerous event ↦ inherent safety; passive safety **D** - **F** - **Pl** *wbudowane bezpieczeństwo* **Sv** *in-byggd säkerhet*

engineered storage *wst* • (in waste management:) protected storage in a facility specially constructed for this purpose ↦ waste management; interim storage; ultimate storage **D** *technische Zwischenlagerung* **F** *stockage technique* **Pl** *składowisko strzeżone* **Sv** *övervakad förvar-ing*

enriched fuel[1] *nf* • nuclear fuel that contains enriched uranium ↦ en-riched uranium; enriched fuel[2]; fissile mate-rial **D** *angereicherter Brennstoff*[1] **F** *com-bustible enrichi*[1] **Pl** *paliwo wzbogacone*[1] **Sv** *anrikat bränsle*

enriched fuel[2] *nf* • nuclear fuel to which fissile nuclides, that were not present earlier, have been added ↦ enriched fuel[1]; fissile material **D** *angere-icherter Brennstoff*[2] **F** *combustible enrichi*[2] **Pl** *paliwo wzbogacone*[2] **Sv** *berikat bränsle*

enriched material *nf; mat* • material in which the content of one or more specific isotopes of a certain initial el-ement has been increased in compar-ison to the naturally occurring con-tent ↦ enriched fuel; enriched uranium; isotope; element **D** *angereichertes Material* **F** *matière enrichie* **Pl** *materiał wzbogacony* **Sv** *anrikat material*

enriched reactor *rty* • nuclear re-actor that uses enriched fuel[1] ↦ en-riched fuel[1]; enriched fuel[2] **D** *Reaktor mit angereichertem Brennstoff* **F** *réacteur à com-bustible enrichi* **Pl** *reaktor na paliwo wzbo-gacone* **Sv** *anrikad reaktor*

enriched uranium[1] *nf; mat* • ura-nium with a higher content of the iso-tope ^{235}U than naturally occurring ↦ enriched material; enriched uranium[2]; en-riched fuel **D** *angereichertes Uran*[1] **F** *ura-nium enrichi*[1] **Pl** *uran wzbogacony*[1] **Sv** *an-rikat uran*

enriched uranium[2] *nf; mat* • ura-nium that contains the isotope ^{233}U ↦ enriched material; enriched uranium[1]; enriched fuel **D** *angereichertes Uran*[2] **F** *uranium enrichi*[2] **Pl** *uran wzbogacony*[2] **Sv** *berikat uran*

enrichment[1] *nf* • **1.** increase of the content of one or several specified isotopes in an element; *enrichment techniques* include: (a) gaseous dif-fusion process, (b) centrifugal pro-cess, (c) *aerodynamic separation pro-cess* in which a mixture of hydrogen and UF_6 is subjected to strong aero-dynamic forces to separate $^{235}UF_6$ from $^{238}UF_6$, (d) electromagnetic sep-aration process, and (e) laser pro-cess **2.** fraction of atoms of a cer-tain isotope in a mixture of isotopes of the same element, when this con-tent is larger than naturally occurring ↦ enrichment[2]; gaseous diffusion process; centrifugal process; electromagnetic separation process; laser process; isotope separation; isotope; element; enrichment factor; enriched material **D** *Anreicherung*[1] **F** *enrichissement*[1] **Pl** *wzbogacenie*[1] **Sv** *anrikning*

enrichment[2] *nf* • (for nuclear fuel:) addition of the fissile material in a form of a not previously present chemical element ↦ enrichment[1]; fissile material **D** *Anreicherung*[2] **F** *enrichissement*[2] **Pl** *wzbogacenie*[2] **Sv** *berikning*

enrichment factor *nf* • ratio of a fraction of atoms of a given isotope in a mixture enriched in this isotope and a fraction of atoms of this isotope in a mixture with the natural composition ↦ isotope; degree of enrichment **D** *Anreicherungsfaktor* **F** *facteur d'enrichissement* **Pl** *współczynnik wzbogacenia* **Sv** *anrikningsfaktor*

enrichment techniques ↦enrichment[1]

enthalpy *th* • thermodynamic state function defined as $H = U + pV$, where H is the e., U is the internal energy, p is the pressure and V is the volume; when heat is added to a system at constant pressure $dQ = dH = C_p dT$, the e. changes by exactly the same amount of heat that is added, where dT - differential increase of temperature and C_p - specific heat at constant pressure ↦ internal energy **D** *Enthalpie* **F** *enthalpie* **Pl** *entalpia* **Sv** *entalpi*

EOC ↦end of cycle

epithermal neutron *rd* • neutron which has kinetic energy higher than that of the thermal neutron but lower than that of the cadmium neutron ↦ neutron energy distribution; thermal neutron **D** *epithermisches Neutron* **F** *neutron épithermique* **Pl** *neutron epitermiczny* **Sv** *epitermisk neutron*

epithermal reactor *rty* • reactor in which a significant number of fissions

is caused by epithermal neutrons ↦ epithermal neutron **D** *epithermischer Reaktor* **F** *réacteur épithermique* **Pl** *reaktor na neutronach epitermicznych; reaktor epitermiczny* **Sv** *epitermisk reeaktor*

EPR *rty* • (acronym for *E*uropean *P*ressurized water *R*eactor:) designed by AREVA, four-loop evolutionary pressurized water reactor configuration with capability of producing over 1600 MWe, employing extensive active safety systems; each circulation loop consists of one steam generator and one circulation pump; EPR is designed to use 17×17 pin array fuel assemblies containing mixed-oxide (PuO_2/UO_2) fuel ↦ pressurized water reactor **D** *EPR* **F** *EPR* **Pl** *EPR* **Sv** *EPR*

EPZ ↦emergency planning zone

equation of state *th* • thermodynamic equation relating state variables which describe the state of matter under given conditions, which in general can be written as $\rho = f(p, T)$, where ρ is the density, p is the pressure and T is the absolute temperature; for an ideal gas, the e.o.s is expressed by the ideal gas law; for real gases and liquids, various formulations are used, e.g., van der Waals equation or multi-parameter equations of state ↦ ideal gas law; van der Waals equation **D** *Zustandsgleichung* **F** *équation d'état* **Pl** *równanie stanu* **Sv** *tillståndsekvation*

equilibrium core *roc* • reactor core in which nuclear fuel is part of an equilibrium cycle ↦ equilibrium cycle **D** *Gleichgewichtskern* **F** *coeur en phase d'équilibre* **Pl** *rdzeń zrównoważony* **Sv** *fortfarighetshärd; jämviktshärd*

equilibrium cycle *nf* • ideal fuel cycle, used in calculation of the fuel cost in a nuclear power plant, in which the newly inserted and removed fuel has, correspondingly, the same masses and compositions as in the preceding cycle ↦ fuel cycle **D** *Gleichgewichtszyklus* **F** *phase d'équilibre* **Pl** *zrównoważony cykl paliwowy* **Sv** *jämviktscykel*

equilibrium time *nf* • (for an isotope separation:) time that is needed to achieve a stationary value of the isotope ratio in each separation step in a cascade ↦ isotope separation; cascade **D** *Gleichgewichtszeit* **F** *temps d'équilibre* **Pl** *czas równowagi* **Sv** *inställningstid*

erg *th* • old unit of energy equal to 10^{-7} J, defined as the amount of work done by a force of one dyne exerted on a distance of one centimetre ↦ joule; dyne **D** *Erg* **F** *erg* **Pl** *erg* **Sv** *erg*

error tolerance *rs* • (in relation to reactor safety, also called *forgivingness*:) level of the human error handling or lack of handling that can be tolerated without jeopardizing the safety of a nuclear reactor ↦ reactor **D** - **F** - **Pl** *tolerancja błędu* **Sv** *feltolerans*

eV ↦*electron volt*

event *rs* • any occurrence unintended by the operator, including operating error, equipment failure or other mishap, and deliberate action on the part of others, the consequences or potential consequences of which are not negligible from the point of view of protection or safety ↦ accident; INES **D** *Ereignis* **F** *événement* **Pl** *zdarzenie* **Sv** *händelse*

excess reactivity *rph* • value of the reactivity that a reactor core at any well-defined state would achieve if all movable control elements were instantaneously removed from the core ↦ control element; reactivity **D** *Überschußreaktivität* **F** *excédent de réactivité* **Pl**

reaktywność nadmierna **Sv** *överskottsreaktivitet*

exchange distillation *nf* • isotope separation process based on the distillation of a chemical compound, leading to a dissociated vapour phase ↦ isotope separation **D** *Austauschdestillation* **F** *distillation avec échange isotopique* **Pl** *destylacja wymienna* **Sv** *utbytesdestillation*

exclusion area *rdp* • zone around a nuclear installation in which certain construction restrictions are valid ↦ nuclear installation **D** *Sperrbereich* **F** *zone d'exclusion* **Pl** *strefa zakazana* **Sv** *skyddszon*

exemption *rdp* • decision of a radiation protection authority that a radioactive material, for example a radioactive waste, can be used or processed, or that a radioactively contaminated area can be accessed without a limitation due to radiation ↦ radioactive waste; radioactive contamination; radioactive material **D** *Freigabe* **F** *exemption* **Pl** *zwolnienie* **Sv** *friklassning*

experimental reactor *rty* • nuclear reactor that is mainly used to obtain physical and experimental data for new reactors and reactor types ↦ research reactor; source reactor **D** *Versuchsreaktor* **F** *réacteur expérimental* **Pl** *reaktor eksperymentalny* **Sv** *experimentreaktor*

exponential assembly *mt* • assembly used to perform exponential experiments ↦ exponential experiment **D** *exponentielle Anordnung* **F** *assemblage exponentiel* **Pl** *zestaw wykładniczy* **Sv** *exponentialuppställning*

exponential decay *rdy* • decay of a radioactive nuclide according to the following relationship: $N(t) = N_0 \cdot e^{-\lambda t}$, where $N(t)$ and N_0 are the number of nuclide's nuclei at time t and 0, respectively, and λ is the decay constant ↦ decay; radioactive decay; decay constant **D** *exponentieller Zerfall* **F** *décrois-*

sance exponentielle **Pl** *rozpad wykładniczy*
Sv *exponentiellt sönderfall*
exponential experiment *mt* • experiment with a sub-critical assembly
of a reactor material and an independent neutron source to determine the
neutronic properties of the material in
the assembly; the adjective "exponential" is motivated because the neutron
flux density with the usual location
of the source gets an exponential distribution in one direction ↦ neutron
source; exponential assembly; neutron flux
density **D** *Exponentialexperiment* **F** *expérience exponentielle* **Pl** *eksperyment wykładniczy* **Sv** *exponentialförsök*
exposure[1] *rd* • act or condition (both
intentional and unintentional) of being subject to irradiation; it can be
termed an external exposure, when
radiation originates from a source outside of a body, or an internal exposure, when the source of radiation is
within the body ↦ dose; radiation **D** *Exposition; Bestrahlung* **F** *radioexposition; irradiation* **Pl** *napromienianie* **Sv** *bestrålning; exponering*
exposure[2] *rdp* • (for X radiation
and gamma radiation, expressed in
C/kg:) sum of all ion charges (of
the same sign) per unit mass of air
that has been achieved when all electrons liberated by photons are completely stopped ↦ ion; electron; photon; X radiation; gamma radiation **D** *Standard-Ionendosisstrahlung* **F** *exposition* **Pl** *ekspozycja* **Sv** *exposition*
external event *rs* • initial event
caused by a natural phenomenon or
human activity outside of an installation, usually including both a fire

or flooding in the installation ↦ event;
internal event; initiating event **D** *externes Ereignis* **F** *événement externe* **Pl** *zdarzenie zewnętrzne* **Sv** *yttre händelse*
external irradiation *rdp* • irradiation of a biological object by a radiation source which is located outside
of the object ↦ exposure; internal irradiation; radiation source **D** *äußere Bestrahlung* **F** *irradiation externe* **Pl** *napromienianie zewnętrzne* **Sv** *extern bestrålning*
external source term ↦*radioactive source term*
extraction cycle *nf* • processing
stage which includes solvent extraction, sometimes scrubbing and stripping ↦ solvent extraction; scrubbing; stripping **D** *Extraktionskreislauf; Extraktionszyklus* **F** *cycle d'extraction* **Pl** *cykl ekstrakcji* **Sv** *extraktionscykel*
extrapolated boundary *rph* • imaginary surface outside of a body on
which the neutron flux density is zero,
when it is extrapolated from flux distribution inside the body ↦ neutron
flux density **D** *extrapolierte Grenze* **F** *limite extrapolée* **Pl** *ekstrapolowana powierzchnia brzegowa* **Sv** *extrapolerad randyta*
extrapolation distance *rph* • distance determined with one-group
model between a material body surface and a point, where the asymptotic neutron flux density would be
zero, if represented with the same
function as inside the body ↦ onegroup model; neutron flux density; linear extrapolation distance **D** *extrapolierte Länge* **F** *longueur extrapolée* **Pl** *długość ekstrapolacji* **Sv** *extrapolationslängd*
extrapolation length ↦*linear extrapolation distance*

Ff

F ↦ *fluorine*

Fahrenheit (temperature) scale
th • temperature scale with 1 °F as
a unit, in which water freezes at 32 °F
and boils at 212 °F; a conversion from
the *F.t.s.* to the Celsius scale is: [°C]
$= ([°F] - 32) \times \frac{5}{9}$, and to the Kelvin
scale: $[K] = ([°F] + 459.67) \times \frac{5}{9}$ ↦ Celsius (temperature) scale; kelvin **D** *Fahren-*
heit Temperaturskala **F** *échelle de tempéra-*
ture en degrés Fahrenheit **Pl** *skala (temper-*
atur) Fahrenheita **Sv** *Fahrenheit-skala*

failed element ↦ *burst can*

fail safe *rs* • (concerning a reactor
engineering system:) which automatically transits into a safe state, if a
failure in the system occurs ↦ re-
actor engineering **D** *folgeschadensicher* **F**
sûr après défaillance **Pl** *bezpieczny w razie*
uszkodzenia **Sv** *säker vid fel*

failure factor *roc* • (concerning a reactor unit, expressed in percent:) factor found as $f = 100 T_f/(T_f + T_g)$,
where T_f - the time during which a
reactor unit is shutdown due to a failure, T_g - the total time during which
the reactor is coupled to the grid ↦ reactor unit **D** *Ausfallfaktor* **F** *facteur de dé-*
faillance **Pl** *wskaźnik awaryjności* **Sv** *felfak-*
tor; tidfelfaktor

Fanning friction factor *th* • friction
factor for a single-phase fluid flow in a
channel defined as $C_f = \tau_w/(\frac{1}{2}\rho U^2)$,
where τ_w - wall shear stress, ρ - fluid
density, U - mean fluid velocity in the
channel ↦ Darcy friction factor **D - F - Pl**
współczynnik oporu Fanninga **Sv** *Fannings*
friktionsfaktor

Farmer diagram *rs* • diagram showing the calculated relationship between the release of ^{131}I from a nuclear power plant and the probability
of such a release per one year of operation; the diagram was used in the past
to characterise a risk for a nuclear reactor accident ↦ iodine **D** *Farmersche*
Kurve **F** *diagramme de Farmer* **Pl** *wykres*
Farmera **Sv** *Farmer-diagram*

fast burst *nf* • (for the nuclear fuel:)
rapidly progressing failure of a fuel
cladding ↦ fuel; cladding **D** *Aufreißen*
des Hüllrohres **F** *rupture brutale de gaine*
Pl *szybkie rozerwanie* **Sv** *snabbrott*

fast fission *rph* • fission caused
by fast neutrons ↦ fission; fast neu-
tron **D** *Schnellspaltung* **F** *fission rapide*
Pl *rozszczepienie neutronami prędkimi* **Sv**
snabbfission

fast-fission factor *rph* • (denoted ϵ:)
factor which takes into account a contribution to the multiplication factor
of a thermal reactor caused by neutrons resulting from fast fissions; the
f.-f.f. is defined as $\epsilon \equiv$ (total number of fission neutrons, from both fast
and thermal fission)/(number of fission neutrons from thermal fissions)
and ranges between 1.02 and 1.30, depending on the moderator material
and the fuel enrichment employed ↦

fast fission; multiplication factor; four-factor formula **D** *Schnellspaltfaktor* **F** *facteur de fission rapide* **Pl** *współczynnik rozszczepienia neutronami prędkimi* **Sv** *snabbfissionsfaktor*

fast neutron *rph* • neutron with the kinetic energy above a certain limit, which in the reactor physics is usually set equal to 0.1 MeV ↦ reactor physics; neutron; MeV **D** *schnelles Neutron* **F** *neutron rapide* **Pl** *neutron prędki* **Sv** *snabb neutron*

fast reactor *rty* • reactor in which fissions are mainly caused by fast neutrons ↦ fission; fast neutron **D** *schneller Reaktor* **F** *réacteur à neutrons rapides* **Pl** *reaktor prędki; reaktor na neutronach prędkich* **Sv** *snabbreaktor*

Fe ↦ *iron*

feed component *nf* • part of the cost of the enriched uranium attributed to the uranium provided to the isotope separation facility ↦ isotope separation; separative work; enriched uranium **D** *Eingangsmaterial-Kosten* **F** *part alimentation* **Pl** *koszt materiału doprowadzonego* **Sv** *kostnad för tillfört material*

feed water *th* • water, mainly as a condensate from the turbine condenser, which is supplied to a reactor or to steam generators ↦ condenser; steam generator; reactor **D** *Speisewasser* **F** *eau d'alimentation* **Pl** *woda zasilająca* **Sv** *matarvatten; mava*

Fermi age equation *rph* • equation that gives a relationship between the slowing-down density and the spatial coordinates: $\nabla^2 q = \frac{\partial q}{\partial \tau}$ where q is the slowing-down density and τ is the age ↦ slowing-down density; age; Fermi age theory **D** *Fermi-Alter-Gleichung* **F** *équation de l'âge de Fermi* **Pl** *równanie wieku Fermiego* **Sv** *åldersekvation*

Fermi age theory *rph* • theory that describes neutron slowing down, in which it is assumed that the slowing down is continuous and takes place in the same way for all neutrons, and that the spatial neutron distribution can be described using the diffusion theory ↦ age; Fermi age equation; diffusion theory **D** *Fermi-Alter-Theorie* **F** *théorie de l'âge de Fermi* **Pl** *teoria wieku Fermiego* **Sv** *åldersteori*

fertile *rph* • (about a nuclide:) which directly or indirectly can be transmuted into a fissile nuclide through a neutron capture ↦ fissile; fertile nuclide; fertile material **D** *brütbar* **F** *fertile* **Pl** *rodny* **Sv** *fertil*

fertile material *rph* • material that contains one or more fertile nuclides ↦ fertile; nuclide **D** *Brutstoff* **F** *matière fertile* **Pl** *materiał paliworodny* **Sv** *fertilt material*

fertile nuclide *rph* • nuclide which is fertile, e.g., ^{232}Th and ^{238}U ↦ fertile; fertile material; nuclide **D** *brütbar Nuklid* **F** *nucléide fertile* **Pl** *nuklid rodny* **Sv** *fertil nuklid*

Fick's approximation ↦ *Fick's law*

Fick's first law of diffusion ↦ *Fick's law*

Fick's law *rph* • (also called *Fick's first law of diffusion*:) physical law saying that the mass flow rate of a substance due to diffusion is proportional to the concentration of the substance, with a proportionality factor called the diffusivity; in the neutron diffusion theory, *F.l.* is applied to describe a relationship between the neutron current density **J** and the neutron flux density ϕ as the following *Fick's approximation*: $\mathbf{J} = -D\nabla\phi$, where D is referred to as the diffusion coefficient ↦ neutron current density; neutron flux density; diffusion coefficient; diffusion theory; nabla **D** *Ficksches*

Gesetz F loi de Fick Pl prawo Ficka Sv Ficks lag

FIFA *rph* • (acronym for *F*issions per *I*nitial *F*issile *A*tom:) total number of fissions which have occurred within a certain amount of fuel, divided by the number of fissile atoms that were initially present in the fuel; since the amount of energy released per fission varies, the relationship between FIFA and the specific burnup is not unique ↦ fissile; burnup fraction; specific burnup; FIMA   *D FIFA-Wert; Spaltstoff-Atomabbrand F FIFA Pl FIFA Sv FIFA*

film badge *rdp* • dosimeter that is based on the influence of the ionising radiation on a photographic emulsion ↦ dosimeter; ionizing radiation   *D Filmdosismesser; Filmdosimeter F dosimètre photographique Pl dawkomierz osobisty Sv filmdosimeter*

film boiling *th* • type of boiling heat transfer during which a layer of vapour completely blankets the heated surface and the liquid does not contact the surface ↦ nucleate boiling; subcooled film boiling; critical heat flux; departure from nucleate boiling   *D Filmsieden; Siedekrise; Siedekrisis F caléfaction Pl wrzenie błonowe Sv filmkokning*

film dosimeter ↦*film badge*

filtered venting *rs* • method to prevent an over-pressure in a containment and release of a radioactive material, e.g., through a safety valve with a filter ↦ containment; radioactive material   *D Druckentlastung über Filter F décompression-filtration (de l'enceinte) Pl redukcja ciśnienia przez filtr Sv filtrerad tryckavlastning*

FIMA *rph* • (acronym for *F*issions per *I*nitial *M*etal *A*tom:) total number of fissions, which have occurred within a certain amount of fuel, divided by the number of fissionable atoms which were initially present in the fuel; since the amount of energy released per fission varies, the relationship between FIMA and the specific burnup is not unique; to avoid ambiguity, the term shouldn't be used for a fuel with non-fissile metallic ingredients ↦ fissionable; burnup fraction; specific burnup; FIFA   *D FIMA-Wert; Schwermetall-Atomabbrand F FIMA Pl FIMA Sv FIMA*

final safety analysis report ↦*safety analysis report*

fine-control element *roc* • control element for small and precise adjustments of the nuclear reactor reactivity ↦ fine-control rod; control element; reactivity   *D Feinsteuerelement F élément de réglage fin Pl element sterowania dokładnego Sv finstyrelement*

fine-control rod *roc* • rod-shaped fine-control element ↦ fine-control element   *D Feinsteuerstab F barre de pilotage Pl pręt sterowania dokładnego Sv finstyrstav*

fine structure *rph* • (in reactor physics:) microscopic variation of a reactor parameter, e.g., the neutron flux density, within a cell in a reactor lattice ↦ reactor physics; neutron flux density; reactor lattice   *D Feinstruktur F structure fine Pl struktura subtelna Sv finstruktur*

finite element method *rph* • numerical method in which the analysed system is divided into elements connected with points, lines or surfaces; in each element the analysed quantity (such as, e.g., the neutron flux density, the displacement or the temperature) is approximated with polynomials; the *f.e.m.* is frequently used in the neutron transport theory as well as in the heat conduction and the solid mechanics analyses ↦ discrete ordinates method; neutron flux density **D** *Methode der finiten Elemente* **F** *méthode d'analyse par éléments* **Pl** *metoda elementów skończonych* **Sv** *finita elementmetoden*

first-collision probability *rph* • probability that a neutron, which starts at a certain point, undergoes a collision in a certain region ↦ neutron **D** *Erststoßwahrscheinlichkeit* **F** *probabilité de première collision* **Pl** *prawdopodobienstwo pierwszej kolizji* **Sv** *sannolikhet för första kollision*

fissible *nap* • adjective used to describe a nuclide or material that can sustain a chain reaction with fast neutrons, but not with slow neutrons ↦ fissile; fissile material; fissionable; slow neutron **D** *spaltbar* **F** *fissionnable* **Pl** *rozszczepialny (na neutronach prędkich)* **Sv** *fissibel*

fissile *nap* • adjective to describe a nuclide that can undergo fission when absorbing a thermal neutron; this term shouldn't be used for nuclides, which have a negligible fission cross section for thermal neutrons, such as ^{238}U ↦ fission cross section; thermal neutron; fissionable **D** *thermisch spaltbar* **F** *fissile* **Pl** *rozszczepialny (na neutronach termicznych)* **Sv** *fissil; termiskt klyvbar*

fissile class *rdp* • class to which package containing fissionable material belongs as far as safety is concerned; *f.c.* governs packaging and transport in such a way that criticality can not occur under foreseeable transportation conditions; packages can be criticality-safe, irrespective of the number and stowage (class I), when the number is limited and irrespective of the stowage (class II), or under other circumstances (class III) ↦ fissionable **D** *nukleare Sicherheitsklasse* **F** *classe fissile* **Pl** *klasa (bezpieczenstwa) ma-*

teriału rozszczepialnego **Sv** *nukleär säkerhetsklass*

fissile material *nap* • material containing ^{233}U, ^{235}U, ^{239}Pu, ^{241}Pu or any combination of these nuclides; as *f.m.* do not count: (i) natural uranium or depleted uranium which is unirradiated; (ii) natural uranium or depleted uranium which has been irradiated in thermal reactors only ↦ fissile; nuclide **D** *Spaltstoff* **F** *matière fissile* **Pl** *materiał rozszczepialny* **Sv** *fissilt material*

fission *nap* • (also called *nuclear fission:*) division of an excited compound nucleus, formed after absorption of a neutron, into two or more lighter nuclei, called fission fragments, followed with emissions of fission neutrons and gamma radiation ↦ fissible; fissionable **D** *Kernspaltung* **F** *fission nucléaire* **Pl** *rozszczepianie* **Sv** *klyvning; fission*

fissionable *nap* • adjective used to describe a nuclide that can be fissioned with fast neutrons, such as, e.g., ^{232}Th, ^{238}U and ^{240}Pu, as well as a fissile nuclide such as ^{235}U ↦ fission; fissible; fissile **D** *spaltbar* **F** *fissionnable* **Pl** *rozszczepialny* **Sv** *klyvbar*

fissionable material *nap* • material that contains fissionable nuclides ↦ fissionable; nuclide **D** *Spaltbarstoff* **F** *matière fissionnable* **Pl** *materiał rozszczepialny* **Sv** *klyvbart material*

fission chamber *mt* • ionization chamber used to monitor the neutron flux (and thus the reactor power) or to provide indication, alarms, and reactor trip signals; the *f.ch.* is filled with argon and contains a fissionable material that undergoes fission due to incident neutrons; ionization in the *f.ch.* is mainly caused by fission fragments; miniaturized *f.ch.* can be tailored for in-core use over any of the ranges likely to be encountered in reactor operation ↦ ionization chamber; fission fragment; neutron **D** *Spaltkammer* **F** *chambre à fission* **Pl** *komora (jonizacyjna) rozszczepieniowa* **Sv** *fissionskammare*

fission counter tube *mt* • counter tube that contains fissionable material, intended for detection of neutrons; ionization is mainly caused by fission fragments produced by the neutron absorption ↦ ionization chamber; fission fragment; neutron absorption; fissionable **D** *Spaltzählrohr* **F** *tube compteur à fission* **Pl** *licznik rozszczepieniowy* **Sv** *fissionsräknerör*

fission cross section *xr* • cross section for the fission nuclear reaction in fissile nuclides; the *f.c.s.* varies with neutron energy and has a $1/v$ region at low neutron energies, followed by a resonance region with many well-defined resonance peaks; at energies in excess of a few keV, the *f.c.s.* decreases with increasing neutron energy ↦ cross section; fission **D** *Wirkungsquerschnitt für Kernspaltung* **F** *section efficace de fission* **Pl** *przekrój czynny na rozszczepienie* **Sv** *fissionstvärsnitt*

fission fragment *nap* • nucleus resulting from a nuclear fission, carrying the kinetic energy from that fission; the highly charged *f.f.* passes through the surrounding medium causing millions of Coulombic ionizations and excitation reactions with the electrons of the medium; within about 10^{-12} s after the fission, the *f.f.* gradually acquires electrons and becomes an electrically neutral atom ↦ fission; fission product; nucleus **D** *Spaltungsbruchstück* **F** *fragment de fission* **Pl** *fragment rozszczepienia* **Sv** *fissionsfragment*

fission gamma radiation *rdy* • gamma radiation emitted during nuclear fission; the *f.g.r.* can be either prompt, when emitted immediately, within less than 0.1 μs after the nuclear fission, or delayed, when emitted later ↦ fission; gamma radiation **D** *Spaltungsgammastrahlung* **F** *rayonnement gamma de fission* **Pl** *rozszczepieniowe promieniowanie gamma* **Sv** *fissionsgammastrålning*

fission gas *nap* • gaseous fission product, such as, e.g., ^{133}Xe, ^{135}Xe, ^{85}Kr, ^{87}Kr and ^{88}Kr ↦ fission product **D** *Spaltgas* **F** *gaz de fission* **Pl** *gazowy produkt rozszczepienia* **Sv** *fissionsgas*

fission gas plenum *nf* • plenum on the top of pellet pile in a fuel rod intended for gaseous fission products, which are released due to the fuel irradiation; the *f.g.p.* is dimensioned to limit the pressure inside the cladding due to fission gases ↦ fission gas **D** *Spaltgasraum* **F** *chambre d'expansion pour gaz de fission* **Pl** *komora na gazowe produkty rozszczepienia* **Sv** *fissionsgasutrymme*

fission neutron *rd* • neutron that is emitted after the nuclear fission, either as the prompt or delayed neutron ↦ fission; prompt neutron; delayed neutron **D** *Spaltneutron* **F** *neutron de fission* **Pl** *neutron rozszczepieniowy* **Sv** *fissionsneutron*

fission poison *nap* • fission product that has a significant cross section for neutron absorption, e.g., ^{135}Xe or ^{149}Sm; the *f.p.* is of concern since it acts as a parasitic neutron absorber that decreases the thermal utilization factor f and introduces negative reactivity into a core ↦ fission product; neutron absorption; thermal utilization factor **D** *Spaltgift* **F** *poison de fission* **Pl** *trucizna rozszczepieniowa* **Sv** *fissionsgift*

fission product *nap* • nuclide produced by a nuclear fission after slowing down and emission of prompt neutrons and gamma rays; several hundred different nuclides can be produced by the fission event or by the subsequent decay of the *f.p.*, usually classified in nine groups: noble gases, I, Cs, Te, Sr, Ru, La, Ba and Ce; selected volatile fission products in a 1000 MWe PWR at the end of a fuel cycle are as follows: ^{134}I ($T_{1/2} = 52.3$ m, $y=0.0718$, $a = 1.9 \times 10^8$ Ci), ^{133}I ($T_{1/2} = 20.8$ m, $y = 0.0676$, $a = 1.7 \times 10^8$ Ci), ^{133}Xe ($T_{1/2} = 5.27$ d, $y = 0.0677$, $a = 1.7 \times 10^8$ Ci), ^{135}Xe ($T_{1/2} = 9.2$ h, $y = 0.0672$, $a = 0.34 \times 10^8$ Ci), where $T_{1/2}$ - radioactive half-life, y - fission yield and a - fission product activity ↦ fission; fission fragment; radioactive half-life; fission yield; fission product activity; nuclide **D** *Spaltprodukt* **F** *produit de fission* **Pl** *produkt rozszczepieniowy* **Sv** *fissionsprodukt*

fission product activity *rdy* • radioactivity of a fission product; the highest activity in a 3560 MWt reactor have the following radionuclides: ^{134}I (2.04×10^8 Ci), ^{133}Xe (1.83×10^8 Ci), ^{133}I (1.83×10^8 Ci), ^{140}Ba (1.72×10^8 Ci), and ^{99}Mo (1.72×10^8 Ci) ↦ radioactivity; fission product **D** *Spaltproduktaktivität* **F** *activité des produits de fission* **Pl** *aktywność produktu rozszczepienia* **Sv** *fissionsproduktaktivitet*

fission spectrum *nap* • energy spectrum of prompt neutrons in a certain fission process, usually expressed by Watt's distribution law; sometimes the term refers to energy spectrum of the prompt gamma radiation ↦ prompt neutron; Watt's distribution law; fission gamma radiation **D** *Spaltspektrum* **F** *spectre de fission* **Pl** *widmo rozszczepieniowe* **Sv** *fissionsspektrum*

fission spike *mat* • displacement spike caused by fission fragments ↦ displacement spike; fission fragment **D** *Spaltungsstörzone* **F** *pointe de fission* **Pl** *rozszczepieniowy szczyt przemieszczeń* **Sv** *fissionsfragmentzon*

fission yield *nap* • fraction of all fissions resulting in a given type of fission fragment ↦ fission fragment; fission **D** *Spaltausbeute* **F** *rendement de fission* **Pl** *wydajność fragmentów rozszczepienia* **Sv** *fissionsutbyte*

fissium *nap; roc* • **1.** artificial mixture of naturally occurring elements that are represented in fission products, intended, as far as the chemical properties are concerned, to mimic the material which is obtained due to fission; the mixture composition depends on the irradiated material and the irradiation conditions to be imitated; **2.** fission products in the mixture of nuclear fuel and fission products obtained after the nuclear fuel has repeatedly been run through a specific fuel cycle with pyrometallurgical processing ↦ fission product; fission; fuel cycle **D** *Fissium* **F** *fissium* **Pl** *fissium* **Sv** *fissium*

flashing *th* • phase change in which the vapour generation occurs solely as a result of a reduction in the system pressure ↦ boiling **D** *Entspannungsverdampfung* **F** *évaporation par déteute* **Pl** *odparowywanie rzutowe* **Sv** *stötkokning*

flow quality ↦ *steam quality*

fluence ↦ *particle fluence*

fluidized bed reactor *rty* • nuclear reactor in which a fuel dispersed in fluid is kept in the reactor core ↦ nuclear reactor; nuclear fuel **D** *Reaktor mit turbulent fluidisiertem Brennstoff; Fließbettreaktor* **F** *réacteur à lit fluidisé* **Pl** *reaktor ze złożem fluidalnym* **Sv** *flytbäddsreaktor; virvelbäddsreaktor*

fluid-poison control *roc* • reactor control through changing the amount or location of a dissolved or suspended nuclear poison ↦ reactor control; nuclear poison **D** *Steuerung durch flüssige Neutronengifte* **F** *commande par poison fluide* **Pl**

sterowanie ciekłą trucizną **Sv** *fluidgiftstyrn-ing*

fluorine *mat* • chemical element denoted F, with atomic number $Z=9$, relative atomic mass $A_r=18.9984032$, density 1.50 g/cm^3, melting point -219.62 °C, boiling point -188.12 °C, crustal average abundance 585 mg/kg and ocean abundance 1.3 mg/L; since *f.* consists of the single isotope ^{19}F, it can be used in the enrichment of uranium using uranium hexafluoride, UF$_6$, since the difference in molecular weights of different molecules of UF$_6$ is only due to the difference in weights of the uranium isotopes $\mapsto$ element **D** *Fluor* **F** *fluor* **Pl** *fluor* **Sv** *fluor*

flux flattening *rph* • obtaining of a more uniform neutron flux density distribution in the nuclear reactor core $\mapsto$ neutron flux density; flux-peaking factor **D** *Abflachung der Flußverteilung* **F** *aplatissment du flux* **Pl** *spłaszczanie strumienia; wyrównywanie strumienia* **Sv** *flödesutjämning*

flux-peaking factor *rph* • (also called *flux-shape factor*:) ratio of the local maximum value of the neutron flux density to its mean value in the reactor core $\mapsto$ neutron flux density; power-peaking factor **D** *Flußformfaktor* **F** *facteur (de forme) de flux* **Pl** *współczynnik rozkładu strumienia* **Sv** *flödesformfaktor*

flux-shape factor $\mapsto$*flux-peaking factor*

flux trap *rph* • region in a moderator material that in a (usually undermoderated) reactor core causes a local increase of the thermal neutron flux density $\mapsto$ neutron flux density; undermoderated **D** *Flußfalle* **F** *piège à flux* **Pl** *pułapka strumienia* **Sv** *flödesfälla*

follower *roc* • extension of a control rod, with or without nuclear fuel, which occupies the empty space cre-ated after the control rod withdrawal $\mapsto$ control rod **D** *Folgestab* **F** *prolongateur* **Pl** *przedłużacz (pręta regulacyjnego)* **Sv** *styrstavsföljare; följare*

forced convection *th* • convection heat transfer, in which the fluid is in motion due to external forces generated by various machines or devices such as pumps, fans or blowers; the *f.c.* plays an important role in evacuation of heat from the core during the normal reactor operation and after the reactor shutdown; the *f.c.* is a much more efficient heat transfer mode than the natural convection, but it requires an active element (a pump or a blower) to provide the coolant circulation $\mapsto$ natural convection **D** *erzwungene Konvektion* **F** *convection forcée* **Pl** *konwekcja wymuszona* **Sv** *påtvingad konvektion*

forgivingness $\mapsto$*error tolerance*

four-factor formula *rph* • expression to calculate the infinite medium multiplication factor k_∞ as a product of four factors: p - resonance escape probability, ϵ - fast-fission factor, η - number of fission neutrons produced per absorption in fuel, f - thermal utilization factor $\mapsto$ infinite multiplication constant; fast-fission factor; resonance escape probability; neutron yield per absorption; thermal utilization factor **D** *Vier-Faktoren-Formel* **F** *formule des quatre facteurs* **Pl** *iloczyn czterech współczynników* **Sv** *fyrfaktorformel*

Fourier's law of heat conduction *th* • physical law stating that the conduction heat flux in a material is proportional to the temperature gradient $\mathbf{q}'' = -\lambda\nabla T$, where $\mathbf{q}''$ is the conduction heat flux vector, λ is a thermal conductivity of the material and T is the temperature $\mapsto$ heat conduction; thermal conductivity; heat transfer; nabla

D *Fouriersches Gesetz* **F** *loi de Fourier* **Pl** *prawo przewodnictwa cieplnego Fouriera* **Sv** *Fouriers lag*

frame monitor *rdp* • frame-shaped radiation monitor for the personnel monitoring when passing through the frame ↦ radiation monitor   **D** *Durchgangsmonitor* **F** *portique de contrôle* **Pl** *portal monitorujący* **Sv** *rammonitor; portalmonitor*

free-air ionizing chamber *rdp* • airfilled ionization chamber in which a certain photon beam passes between electrodes in such a way that neither photons nor generated electrons hit electrodes, which makes the region used for calculation of exposure well defined; the *f-a.i.c.* is mainly used as a standard ionization chamber ↦ ionization chamber; exposure **D** *FreiluftIonisationskammer* **F** *chambre d'ionisation à air libre* **Pl** *komora jonizacyjna wzorcowa* **Sv** *öppen luftkammare*

free convection ↦*natural convection*

free-gas model *rph* • model to calculate the scattering kernel for thermal neutrons, in which it is assumed that the moderating nuclei are free particles and their chemical bindings are ignored ↦ scattering kernel; thermal neutrons **D** *Gasmodell* **F** *modèle des gaz libres* **Pl** *model gazu swobodnego* **Sv** *fri gasmodell*

freestanding cladding *nf* • cladding constructed in such a way that it resists coolant over-pressure without any support from the nuclear fuel ↦ cladding; collapsible cladding **D** *freistehende Hülle* **F** *gaine résistante* **Pl** *koszulka samonośna* **Sv** *fristående kapsel*

fresh fuel *nf* • new or unirradiated fuel, including fuel fabricated from fissionable material recovered by reprocessing previously irradiated fuel ↦ fissionable material; nuclear fuel **D** *frischer*

Brennstoff **F** *combustible neuf* **Pl** *paliwo świeże* **Sv** *färskt bränsle*

front-end of the fuel cycle *nf* • initial stage of a nuclear fuel cycle including the evaluation of uranium ore deposits, uranium ore mining, production of UF_6, enrichment of uranium and nuclear fuel manufacturing ↦ back-end of the fuel cycle; fuel cycle; nuclear fuel **D** *Anfang des Brennstoffkreislaufs* **F** *partie initiale du cycle du combustible* **Pl** *etap wstępny cyklu paliwowego* **Sv** *begynnelsesteg[1]*

fuel ↦*nuclear fuel*

fuel assembly *nf* • reactor component that contains fuel elements and which is operated as one entity when it is placed in or removed out from a reactor core ↦ fuel element; reactor core **D** *Brennelementbündel[1]* **F** *assemblage combustible* **Pl** *kaseta paliwowa; sekcja paliwowa; zespół paliwowy; zestaw paliwowy* **Sv** *bränslepatron*

fuel batch ↦*fuel charge*

fuel box *nf* • canister with a quadratic cross section containing fuel channel in a boiling water reactor ↦ fuel channel; boiling water reactor **D** *Brennelementkasten* **F** *chemise d'un assemblage combustible* **Pl** *powłoka kasety paliwowej* **Sv** *bränslebox*

fuel bundle *nf* • bundle of fuel rods, usually organized in parallel to each other ↦ fuel rod **D** *Brennelementbündel[2]* **F** *grappe de combustible* **Pl** *wiązka paliwowa* **Sv** *bränsleknippe; stavknippe*

fuel burnout *th* • local damage of a fuel element due to deterioration of the heat transfer intensity when the local heat flux exceeds the critical heat flux value ↦ critical heat flux **D** *Durchbrennen* **F** *conséquence destructrice de la caléfaction* **Pl** *przepał elementu paliwowego* **Sv** *sönderbränning*

fuel channel *nf* • hole or a pipe

through a moderator which is dedicated to contain fuel assembly and through which a coolant is circulating ↦ moderator; fuel assembly; coolant **D** *Brennelementkanal* **F** *canal de combustible* **Pl** *kanał paliwowy* **Sv** *bränslekanal*

fuel charge *nf* • amount of nuclear fuel in a nuclear reactor; it can refer to the whole reactor core or to this part of the core which is exchanged during refuelling ↦ fuel inventory; once-through charge; reactor core; refuelling **D** *Brennstoffladung; Ladung* **F** *charge de combustible; charge* **Pl** *wsad (paliwowy)* **Sv** *bränsleladdning; laddning; bränslesats*

fuel charging machine *nf* • (also called *fuel discharging machine*:) machine which is used in a nuclear reactor to insert and remove nuclear fuel from the reactor core ↦ nuclear fuel; nuclear reactor **D** *Lademaschine* **F** *machine de chargement* **Pl** *maszyna załadowcza* **Sv** *laddmaskin*

fuel cluster ↦*fuel bundle*

fuel consumption charge *nf* • (also called *fuel depletion charge*:) cost of burnup and loss of fuel during treatment after usage in a nuclear reactor; the *f.c.c.* includes depreciation due to a changed isotope content and takes into account the value of plutonium ↦ nuclear fuel; nuclear reactor; plutonium credit **D** *Brennstoffverbrauchskosten* **F** *coût de la consommation du combustible* **Pl** *koszt zużycia paliwa* **Sv** *bränsleförbrukningskostnad*

fuel control *roc* • reactor control through a change of nuclear fuel properties, position or amount ↦ reactor control[2]; nuclear fuel; moderator control **D** *Steuerung durch Brennstoff* **F** *commande par le combustible; surveillance du combustible* **Pl** *sterowanie (reaktora) paliwem* **Sv** *bränslestyrning*

fuel-cooling installation ↦*cooling pond*

fuel cycle *nf* • series of steps that nuclear fuel undergoes, including the front end, the usage in a reactor and the back end ↦ front-end of the fuel cycle; back-end of the fuel cycle; nuclear fuel **D** *Brennstoffkreislauf* **F** *cycle du combustible* **Pl** *cykl paliwowy* **Sv** *bränslecykel; bränslegång*

fuel densification *nf* • increase of density of the ceramic fuel; the *f.d.* takes place due to the high temperature and radiation in the nuclear reactor and results in the volume reduction ↦ ceramic fuel **D** *Brennstoffverdichtung* **F** *densification du combustible* **Pl** *zagęszczenie paliwa* **Sv** *bränsleförtätning*

fuel depletion charge ↦*fuel consumption charge*

fuel discharging machine ↦*fuel charging machine*

fuel element *nf* • smallest, usually canned, construction element containing nuclear fuel as the major component and which is dedicated to a nuclear reactor; the *f.e.* can have various shapes, such as rod, plate or ball ↦ dummy element; fuel rod; fuel assembly **D** *Brennelement* **F** *élément combustible* **Pl** *element paliwowy* **Sv** *bränsleelement*

fuel gap ↦*axial fuel gap*

fuel handling *nf* • handling of fuel in connection to, e.g., refuelling or ultimate waste disposal ↦ refuelling; ultimate waste disposal; fuel **D** *Brennstoffhandhabung* **F** *manutention du combustible* **Pl** *obsługa paliwa* **Sv** *bränslehantering*[1]

fuel inventory *nf* • total amount of nuclear fuel in a reactor, within a system of nuclear reactors or within a fuel cycle ↦ nuclear fuel; fuel cycle; fuel charge; refuelling; reactor core **D** *Brennstoffeinsatz; Brennstoffinventar* **F** *inventaire de* *combustible* **Pl** *zasób paliwa* **Sv** *bränsleinnehåll*

fuel management *nf* • planning and execution of measures to secure access to the nuclear fuel, its best usage in a reactor, and following it, the fuel reprocessing ↦ in-core fuel management; fuel reprocessing **D** *Brennstoffmanagement* **F** *gestion du combustible* **Pl** *gospodarka paliwem jądrowym* **Sv** *bränslehantering*[2]

fuel pellet *nf* • fuel body, usually having shape of a short cylinder, intended to be stacked in a cladding to form a fuel element ↦ fuel; cladding; fuel element **D** *Brennstofftablette* **F** *pastille de combustible* **Pl** *pastylka paliwowa* **Sv** *bränslekuts*

fuel processing plant *nf* • installation designed to reprocess fuel ↦ fuel reprocessing **D** - **F** *installation de retraitement du combustible* **Pl** *zakład przeróbki wypalonego paliwa* **Sv** *bränsle upparbetningsanläggning*

fuel rating *th* • ratio of the total thermal power of a nuclear reactor to the initial mass of fissile and fertile nuclides; sometimes the total initial mass of fuel charge is used in the ratio; the *f.r.* is usually expressed in megawatts per tonne ↦ thermal power; fissile; fertile; nuclide; fuel charge; specific power **D** *spezifische Leistung* **F** *puissance spécifique* **Pl** *moc właściwa* **Sv** *bränslebelastning*

fuel reconstitution *nf* • reconstruction of partly burned-up fuel assembly by replacement of one or more fuel rods; the *f.r.* can be performed to improve the reactivity and fuel economy after a certain period of reactor operation ↦ fuel assembly; fuel rod; reactivity **D** *Brennstoffumordnung* **F** *reconsti-*

tution du combustible **Pl** *odbudowa paliwa; rekonstrukcja paliwa* **Sv** *bränsleombyggnad*

fuel reprocessing *nf* • mechanical or chemical treatment of spent fuel to separate fission products and to extract fissile and fertile material ↦ spent fuel; fission product; fissile; fertile **D** *Brennstoffaufarbeitung* **F** *retraitement du combustible; retraitement* **Pl** *przeróbka paliwa wypalonego* **Sv** *bränsleupparbetning; upparbetning*

fuel rod *nf* • rod-shaped fuel element ↦ fuel element **D** *Brennstab* **F** *barre de combustible; crayon combustible* **Pl** *pręt paliwowy* **Sv** *bränslestav*

fuel specific enthalpy *th* • fuel enthalpy per unit mass ↦ enthalpy **D** - **F** - **Pl** *entalpia właściwa paliwa* **Sv** -

fuel-use charge *nf* • rent cost of nuclear fuel ↦ nuclear fuel **D** *Brennstoffkosten* **F** *frais de location de combustible* **Pl** *opłata lokacyjna paliwa* **Sv** *bränslehyra*

Fukushima Daiichi nuclear accident *rs* • nuclear accident that took place in Japan, in Fukushima Daiichi nuclear power plant, on March 4, 2011, in which, following a major earthquake, a 15-metre tsunami disabled the power supply and cooling of three reactors, causing a nuclear accident; all three cores largely melted in the first three days; the accident was rated 7 on the INES scale, due to high radioactive releases over days 4 to 6, totalling to some 940 PBq (^{131}I equivalent) ↦ nuclear accident; INES; Chernobyl nuclear accident; Three Mile Island nuclear accident; tsunami **D** *Nuklearkatastrophe von Fukushima* **F** *accident nucléaire de Fukushima* **Pl** *awaria w Fukushimie* **Sv** *Fukushima-olyckan*

Gg

gamma radiation *rd* • (also called *gamma rays*:) electromagnetic radiation that is emitted due to nuclear processes, such as gamma decay, or as a result of annihilation ↦ alpha radiation; beta radiation **D** *Gammastrahlung* **F** *rayonnement gamma; rayons gamma* **Pl** *promieniowanie gamma* **Sv** *gammastrålning*

gamma radiography *mt* • radiography with gamma radiation ↦ radiography[1]; gamma radiation **D** *Gammaradiographie* **F** *radiographie gamma; gammagraphie* **Pl** *radiografia gamma; gammagrafia* **Sv** *gammaradiografi*

gamma rays ↦*gamma radiation*

gamma scanning *mt* • measurement of the intensity distribution of gamma radiation, e.g., to determine the axial distribution of the burnup in a fuel rod ↦ gamma radiation; burnup; fuel rod **D** *Gamma-Scanning* **F** *exploration gamma* **Pl** *skanowanie gamma* **Sv** *gammaavsökning*

gamma source *rd* • radiation source emitting gamma radiation ↦ radiation source; burnup; gamma radiation **D** *Gammaquelle* **F** *source gamma* **Pl** *źródło promieniowania gamma* **Sv** *gammakälla*

gamma thermometer *mt* • detector of gamma radiation based on the thermocouple principle; the output signal of the *g.t.* is a measure of the heat generation rate in the thermometer body due to the gamma radiation absorption; the *g.t.* is frequently used in the travelling in-core probe system ↦ gamma radiation; travelling in-core probe system; local power range monitor **D** *Gammathermometer* **F** *thermométre gamma* **Pl** *termometr gamma* **Sv** *gammatermometer*

gap ↦*clearance*

gas constant ↦*universal gas constant*

gas-cooled fast reactor *rty* • (abbreviated *GFR:*) generation IV reactor with the fast neutron spectrum in which a gas, such as carbon dioxide or helium, is used as the coolant ↦ generation IV reactor **D** *schneller gasgekühlter Reaktor* **F** *réacteur rapide refroidi par gaz* **Pl** *reaktor prędki chłodzony gazem* **Sv** *gaskyld snabb reaktor*

gas-cooled reactor *rty* • (abbreviated *GCR:*) nuclear reactor in which a gas, such as carbon dioxide or helium, is used as the coolant ↦ reactor; coolant **D** *gasgekühlter Reaktor* **F** *réacteur refroidi par gaz* **Pl** *reaktor chłodzony gazem* **Sv** *gaskyld reaktor*

gaseous diffusion process *nf* • process of the isotope separation which is based on the dependence of the diffusion through a porous membrane on the mass of the diffusing particles ↦ diffusion barrier; isotope separation **D** *Gasdiffusionsprozess* **F** *diffusion gazeuse* **Pl** *proces dyfuzji gazowów* **Sv** *gasdiffusionsprocess*

GCR ↦*gas-cooled reactor*

Ge ↦*germanium*

Geiger counter *mt* • one of the oldest, gas-filled and based on ionization, radiation detector types introduced by Geiger and Mueller in 1928; the *G.c.* is commonly referred to as the *Geiger-Mueller counter*, *G-M counter* or *Geiger tube* ↦ counter tube **D** Geiger-Müller-Zählrohr **F** *tube compteur de Geiger-Müller; compteur Geiger* **Pl** *licznik Geigera-Müllera* **Sv** *GM-rör; Geiger-Müller-rör*

Geiger-Mueller counter ↦*Geiger counter*

Geiger tube ↦*Geiger counter*

general area emergency ↦*reference level for emergency action*

general emergency *rdp* ● highest reference level for emergency action, which is announced after a release of radioactive material or when a probability of such release is very high and requires implementation of safety precautions; as a consequence of the *g.e.* announcement, the preparedness organisation is activated and the public in the precautionary action zone receives an alarm ↦ reference level for emergency action; emergency planning zone **D** *Katastrophenalarm* **F** *alarme d'urgence; siréne d'alerte générale* **Pl** *alarm generalny* **Sv** *haverilarm*

generation IV reactor *rty* ● reactor concept currently under development with envisioned significant improvements over current reactors in economy, safety, waste reduction and non-proliferation; to meet the goals, six concepts are under development by the international community (Generation IV International Forum - GIF): sodium-cooled fast reactor (SFR), gas-cooled fast reactor (GFR), lead-cooled fast reactor (LFR), supercritical water-cooled reactor (SCWR), very high temperature reactor (VHTR) and molten salt reactor (MSR) ↦ sodium-cooled fast reactor; gas-cooled fast reactor; lead-cooled fast reactor; supercritical water-cooled reac-

tor; very high temperature reactor; molten salt reactor; GIF **D** - **F** *génération IV réacteur* **Pl** *reaktor czwartej generacji* **Sv** *fjärde generationens reaktor*

generation time *rph* ● average time between a release of a neutron due to fission and the new fission caused by the neutron ↦ fission; neutron; neutron lifetime; neutron cycle **D** *Generationsdauer* **F** *temps de génération* **Pl** *czas życia pokolenia* **Sv** *generationstid*

genetic effect of radiation *rdp* ● (also called *hereditary effect of radiation:*) changes in heredity due to a damage to the genetic material in germ cells, caused by ionizing radiation; very little information is available on the *g.e.o.r.* in humans and estimations of risks are based on extrapolation of studies in other mammals; however, the evidence from animal studies has left no doubt that heritable mutational effects in humans are possible ↦ ionizing radiation; somatic effect of radiation; radiation damage; radiation injury **D** *genetische Strahlenwirkung* **F** *effets génétiques des rayonnements* **Pl** *genetyczne skutki promieniowania* **Sv** *genetisk strålningsverkan*

geological repository *wst* ● facility for the ultimate storage of radioactive waste in rock ↦ repository; radioactive waste **D** *geologisches Endlager* **F** *dépôt géologique* **Pl** *składowisko geologiczne* **Sv** *geologiskt slutförvar*

geometrically safe *rph* ● (for a system which contains fissile material:) constructed in such a way that a spatial composition or form of the system's components prohibits the occurrence of a self-sustained chain reaction ↦ fissile material; critical mass; nuclear chain-reaction; bird cage **D** *geometrisch sicher* **F** *géométriquement sûr* **Pl**

geometrycznie bezpieczny **Sv** *geometriskt säker*

geometric buckling *rph* • (denoted B_g^2:) parameter that depends on the shape of a body and the body dimensions; the *g.b.* is a measure of the neutron leakage from the body and is equal to the square of the first fundamental mode of the following differential equation: $\nabla^2\psi(r) + B^2\psi(r) = 0$, which describes the spatial neutron flux distribution in the body at any given time; for a finite cylindrical core with extrapolated radius $\tilde{R}$ and height $\tilde{H}$, the *g.b.* is as follows: $B_g^2 = \left(\frac{2.405}{\tilde{R}}\right)^2 + \left(\frac{\pi}{\tilde{H}}\right)^2 \mapsto$ material buckling **D** *geometrische Flussdichtewölbung* **F** *laplacien géométrique* **Pl** *parametr geometryczny reaktora* **Sv** *geometrisk buktighet*

germanium *mat* • chemical element denoted Ge, with atomic number $Z{=}32$, relative atomic mass $A_r{=}72.610$, density 5.323 g/cm^3, melting point 938.25 °C, boiling point 2833 °C, crustal average abundance 1.5 mg/kg and ocean abundance 5×10^{-5} mg/L; *g.* is used in gamma-ray detectors $\mapsto$ element; germanium detector **D** *Germanium* **F** *germanium* **Pl** *german* **Sv** *germanium*

germanium detector *mt* • semiconductor detector with germanium as the main component, mainly used as the gamma spectrometer where high resolution is required $\mapsto$ germanium; semiconductor detector **D** *Germaniumdetektor* **F** *détecteur à germanium* **Pl** *detektor germanowy* **Sv** *germaniumdetektor*

GeV $\mapsto$*electron volt*

GFR $\mapsto$*gas-cooled fast reactor*

GIF *gnt* • (acronym for *G*eneration IV *I*nternational *F*orum:) founded in 2001, co-operative international endeavour which was set up to carry out the research and development needed to establish the feasibility and performance capabilities of the next-generation nuclear energy systems; currently active members of the GIF include: Australia, Canada, People's Republic of China, the European Atomic Energy Community (Euratom), France, Japan, Republic of Korea, Russian Federation, Republic of South Africa, Switzerland, the United Kingdom and the United States; the non-active members are Argentina and Brazil $\mapsto$ generation IV reactor **D** *GIF* **F** *GIF* **Pl** *GIF* **Sv** *GIF*

global criticality *rph* • criticality achieved in the entire reactor core $\mapsto$ criticality; local criticality; reactor core **D** *globale Kritikalität* **F** *criticité globale* **Pl** *krytyczność globalna* **Sv** *global kriticitet*

glove box *rdp* • airtight container fitted with windows and wall-attached gloves, in which toxic or radioactive materials can be viewed and managed $\mapsto$ radioactive material **D** *Handschuhkasten* **F** *boîte à gants* **Pl** *komora rękawicowa* **Sv** *handskbox*

G-M counter $\mapsto$*Geiger counter*

graphite $\mapsto$*carbon*

graphite reactor *rty* • nuclear reactor in which graphite is used as a moderator $\mapsto$ nuclear reactor; moderator; graphite; advanced gas-cooled reactor; RBMK reactor **D** *graphitmoderierter Reaktor* **F** *réacteur modéré au graphite* **Pl** *reaktor grafitowy* **Sv** *grafitreaktor*

Grashof number *th* • non-dimensional number used in natural convection analyses defined as $Gr = \frac{g\beta\Delta T L^3}{\nu^2}$, where g is the gravity acceleration, β is the volume expansion coefficient at constant pressure, ΔT is the temperature difference, L is the characteristic length and ν is the kinematic viscosity ↦ natural convection **D** *Grashof-Zahl* **F** *nombre de Grashof* **Pl** *liczba Grashofa* **Sv** *Grashofs tal*

gray[1] *nap* • (concerning material or device in nuclear engineering:) which absorbs a significant fraction, but not all neutrons of a certain energy, incident to this material or device ↦ absorption; black **D** *grau* **F** *gris* **Pl** *częściowo pochłaniający; szary* **Sv** *delabsorberande*

gray[2] *rdp* • (denoted Gy:) unit of the absorbed dose and kerma, where 1 Gy = 1 J/kg; the *g.* has replaced the old unit of absorbed dose, the rad ↦ absorbed dose; kerma; rad **D** *Gray* **F** *gray* **Pl** *grej* **Sv** *gray*

greenhouse effect *bph* • process due to which radiation from a planet's atmosphere, containing greenhouse gases, warms the planet's surface to a temperature above what it would be without its atmosphere; Earth's natural greenhouse effect is necessary to support life; however, excessive greenhouse effect caused by human activities, such as burning of fossil fuels, can lead to the global warming ↦ radiation **D** *Treibhauseffekt* **F** *effet se serre* **Pl** *efekt cieplarniany* **Sv** *växthuseffekt*

gross power *roc* • electrical power of a reactor unit or a nuclear power plant calculated on output of the electric generator ↦ net power; reactor unit; nuclear power plant **D** *Bruttoleistung* **F** *puissance brute* **Pl** *moc brutto* **Sv** *bruttoeffekt*

gross production *roc* • electrical energy produced by a reactor unit or a nuclear power plant calculated on output of the electric generator ↦ net production; reactor unit; nuclear power plant **D** *Bruttoproduktion* **F** *production brute* **Pl** *produkcja energii brutto* **Sv** *bruttoproduktion*

group cross section *xr* • weighted mean cross section in a neutron energy group ↦ cross section; neutron energy group **D** *Gruppenquerschnitt* **F** *section efficace de groupe* **Pl** *grupowy przekrój czynny* **Sv** *grupptvärsnitt*

group removal cross section *xr* • (within reactor physics:) group cross section which describes the disappearance of neutrons from a neutron energy group due to all occurring processes ↦ group cross section; neutron energy group **D** *Gruppenverlustquerschnitt* **F** *section efficace d'extraction de groupe* **Pl** *przekrój czynny usunięcia z grupy* **Sv** *svinntvärsnitt*[1]

group transfer scattering cross section *xr* • group cross section, characteristic for a neutron energy structure, which corresponds to transfer of neutrons through scattering from a certain energy group to another one; the *g.t.s.c.s.* is an element of the corresponding scattering matrix for group transfer scattering ↦ group cross section; scattering cross section **D** *Gruppenübergangsquerschnitt* **F** *section efficace de transfert de groupe par diffusion* **Pl** *przekrój czynny przeniesienia z grupy do grupy* **Sv** *överföringstvärsnitt*

Hh

H ↦ *hydrogen*

hafnium *mat* • chemical element denoted Hf, with atomic number $Z=72$, relative atomic mass $A_r=178.49$, density 13.3 g/cm^3, melting point 2233 °C, boiling point 4603 °C, crustal average abundance 3.0 mg/kg and ocean abundance 7×10^{-6} mg/L; *h.* is a metallic element with a high neutron absorption cross-section and is a suitable material for control rods ↦ element; control rod **D** *Hafnium* **F** *hafnium* **Pl** *hafn* **Sv** *hafnium*

half-life ↦ *radioactive half-life*

Haling calculation *rph* • iterative procedure to determine the unique three-dimensional power distribution that does not change when a small increase in the specific burnup takes place at the end of an operating period with all control rods withdrawn in a boiling water reactor ↦ specific burnup; operating period; control rod; boiling water reactor **D** *Berechnung der Haling-Verteilung* **F** *calculation Haling* **Pl** *procedura Halinga* **Sv** *Haling-beräkning*

Hankel functions ↦ *Bessel functions of the third kind*

He ↦ *helium*

head-end *nf* • (for fuel reprocessing:) initial processing of spent fuel to a physical and chemical form that is proper for continued reprocessing ↦ fuel reprocessing; spent fuel **D** *Vorbehand-* lung; *Brennstoffaufschluss* **F** *traitement initial* **Pl** *obróbka wstępna* **Sv** *begynnelsesteg^2*

health physics ↦ *radiation* protection

heat conduction *th* • thermal energy transfer mechanism in a solid (the only one), fluid or gaseous body based on transmission of kinetic energy between particles; the necessary condition for *h.c.* is a temperature difference between different points in the body; *h.c.* follows Fourier's law of heat conduction ↦ thermal energy; Fourier's law of heat conduction; convection **D** *Wärmeleitung; Konduktion* **F** *conduction thermique* **Pl** *przewodzenie ciepła* **Sv** *konduktion*

heat-electric station *gnt* • installation that provides both the heat (for example, for district heating) and the electricity ↦ nuclear power plant **D** *Heizkraftwerk* **F** *centrale thermique (électrique)* **Pl** *elektrociepłownia* **Sv** *kraftvärmeverk*

heat flux *th* • (usually denoted q'':) rate of thermal energy transfer through a given surface per unit time ↦ heat transfer; thermal energy **D** *Wärmestromdichte* **F** *flux thermique surfacique* **Pl** *strumień cieplny* **Sv** *värmeflöde*

heat of vaporization *th* • heat required to change the phase of a substance from a saturated liquid to a saturated vapour state ↦ boiling **D** *Verdampfungswärme* **F** *énergie de vaporisa-*

tion *Pl ciepło parowania; ciepło utajone* **Sv** *ångbildningsvärme*

heat transfer *th* • exchange of thermal energy between physical systems; the most important *h.t.* mechanisms include the heat conduction (governed by Fourier's law), the heat convection (governed by Newton's law of cooling) and the radiation (governed by the Stefan-Boltzmann law) ↦ thermal energy; Fourier's law of heat conduction; Newton's law of cooling; Stefan-Boltzmann law; heat transfer coefficient **D** *Wärmeübertraugung* **F** *transfert thermique* **Pl** *wymiana ciepła* **Sv** *värmeöverföring*

heat transfer coefficient *th* • physical quantity defined as a ratio of the heat flux transferred from a solid wall to fluid, and the difference between the wall surface temperature and the fluid bulk temperature: $h = q''/(T_w - T_b)$, where q'' - heat flux, T_w - solid surface temperature, T_b - fluid bulk temperature (this relationship is frequently referred to as *Newton's law of cooling*); *h.t.c.* depends on flow conditions (laminar or turbulent), type of flow (natural circulation or forced circulation), heat transfer type (single-phase heat transfer, boiling heat transfer) and type of solid surface geometry; only for laminar single-phase flow in simple channels, the *h.t.c.* can be determined analytically; in all other situations the coefficient is obtained from a proper correlation valid for the specific conditions; examples of such correlations include: Dittus-Boelter and Colburn correlations (for single-phase turbulent heat transfer in pipes), Chen correlation (for boiling convective heat transfer), Weissman and Markoczy correlations (for single-phase turbulent heat transfer in rod bundles) ↦ Col-

burn correlation; Dittus-Boelter correlation; Markoczy correlation; Weisman correlation **D** *Wärmeübergangskoeffizient* **F** *coefficient de transfert thermique* **Pl** *współczynnik wymiany ciepła* **Sv** *värmeflödeskoefficient*

heavy hydrogen ↦*deuterium*

heavy water *mat* • deuterium oxide, D_2O, which is water in which hydrogen is replaced with deuterium; the *h.w.* is obtained by separation of the deuterated water HDO from the ordinary water, for example, by successive distillation, and then further processing it either physically or chemically ↦ deuterium; deuterated water; heavy-water reactor **D** *schweres Wasser* **F** *eau lourde* **Pl** *ciężka woda* **Sv** *tungt vatten; tungvatten*

heavy-water reactor *rty* • heavy-water moderated nuclear reactor such as, e.g., CANDU, known generically as a pressurized heavy-water reactor ↦ heavy water; moderator; CANDU; PHWR **D** *schwerwassermoderierter Reaktor* **F** *réacteur á eau lourde* **Pl** *reaktor z ciężką wodą; reaktor ciężkowodny* **Sv** *tungvattenreaktor*

helium *mat* • chemical element denoted He, with atomic number $Z=2$, relative atomic mass $A_r=4.002602$, density 0.124901 g/cm^3, melting point -272.2 °C, boiling point -268.93 °C, crustal average abundance 0.008 mg/kg and ocean abundance 7×10^{-6} mg/L; *h.* is used as a coolant in high-temperature gas-cooled reactors ↦ element; coolant; high-temperature gas-cooled reactor **D** *Helium* **F** *hélium* **Pl** *hel* **Sv** *helium*

helium counter tub *mt* • counter tube containing ^{3}He, designed to detect neutrons ↦ helium; counter tube; radiation detector; ionization chamber; neutron **D** *Helium-Zählrohr* **F** *tube compteur á* *hélium* **Pl** *licznik helowy* **Sv** *heliumräknerör; heliumrör*

hereditary effect of radiation
↦*genetic effect of radiation*

heterogeneous reactor *rty* • nuclear reactor in which the core material is distributed in such a way that the neutronic properties of the reactor cannot be described with a satisfactory accuracy if the material is assumed to be homogeneously distributed in the core ↦ homogeneous reactor **D** *heterogener Reaktor* **F** *réacteur hétérogène* **Pl** *reaktor niejednorodny* **Sv** *heterogen reaktor*

HEU ↦*highly enriched uranium*

Hf ↦*hafnium*

high-level waste *wst* • radioactive waste with so high radioactivity that it has to be both shielded and cooled ↦ radioactive waste; radioactivity **D** *hochaktiver Abfall* **F** *déchet de haute activité; déchet d'activité élevée* **Pl** *wysokoaktywny odpad promieniotwórczy* **Sv** *högaktivt avfall*

highly enriched uranium *mat* • (abbreviated *HEU:*) uranium in which the fraction of ^{235}U is higher than 20% ↦ enriched uranium; depleted uranium **D** *hochangereichertes Uran* **F** *uranium hautement enrichi* **Pl** *uran wysokowzbogacony* **Sv** *höganrikat uran*

high-temperature gas-cooled reactor *rty* • (abbreviated *HTGR:*) reactor cooled with helium under high pressure, fuelled with enriched uranium and moderated by graphite; there are two main types of HTGRs: pebble bed reactors (PBR) and prismatic block reactors (PMR); the prismatic block reactor refers to a prismatic block core configuration, in which hexagonal graphite blocks are stacked to fit in a cylindrical pressure vessel; the pebble bed reactor (PBR) design consists of fuel in the form of pebbles, stacked together in a cylindrical pressure vessel; both reactors may have the fuel stacked in an annulus region with a graphite center spire, depending on the design and desired reactor power; the fuel used in HTGRs is coated fuel particles, such as TRISO fuel particles ↦ graphite reactor; ceramic fuel; coated particle; TRISO; pebble-bed reactor **D** *Hochtemperaturreaktor* **F** *réacteur du type graphite-gaz à haut température; réacteur du type HTGR* **Pl** *reaktor wysokotemperaturowy chłodzony gazem* **Sv** *högtemperaturreaktor*[1]*; HTGR*

Hilborn detector ↦*collectron*

hold-up *nf* • amount of material that at the same time and during a steady-state condition is treated in an isotope separation facility or a part thereof ↦ isotope separation **D** *Materialeinsatz* **F** *charge en oeuvre* **Pl** *ilość materiału (związanego w procesie separacji izotopów)* **Sv** *upplagring*

homogeneous flow *th* • multiphase flow model in which all phases or components are assumed to be mixed at a molecular level ↦ two-phase flow **D** - **F** *écoulement homogenique* **Pl** *przepływ jednorodny* **Sv** *homogenströmning*

homogeneous reactor *rty* • nuclear reactor in which core materials are distributed in such a way that the neutron-physical properties can be described with satisfactory accuracy if the materials are assumed to be homogeneously distributed in the reactor core ↦ heterogeneous reactor **D** *homogener Reaktor* **F** *réacteur homogène* **Pl** *reaktor jednorodny* **Sv** *homogen reaktor*

hot *rdp* • (adjective:) highly radioactive ↦ hot cell **D** *heiß* **F** *chaud* **Pl** *wysokoaktywny* **Sv** *högaktiv*

hot cell *rdp* • highly shielded tight casing with lead-glass window, in which highly radioactive material can be processed (using automatic or remotely controlled tools) and stored ↦ hot; hot testing *D heiße Zelle F cellule de haute activité Pl komora wysokoaktywna; komora gorąca Sv högaktiv cell*

hot channel ↦*hot-channel factor*

hot-channel factor *th* • ratio of the highest enthalpy change or the highest linear power in the hot channel, and the average value of the corresponding parameter in the core; here the *hot channel* is defined to be that coolant channel in which the point of maximum linear power (also called the *hot spot* point) occurs; the *h.-ch.f.* fall generally into two main categories: the nuclear factors (related to the neutronic aspects of the core design) and the engineering factors (related to variations and uncertainties of core component dimensions) ↦ coolant channel; power-peaking factor *D Heißkanalfaktor F facteur de canal chaud Pl współczynnik kanału gorącego Sv hetkanalfaktor*

hot laboratory *rdp* • laboratory for operations with significant amounts of radioactive materials ↦ hot; hot cell *D heißes Laboratorium; radioaktives Laboratorium F laboratoire de haute activité Pl laboratorium wysokoaktywne Sv högaktivt laboratorium*

hot spot ↦*hot-channel factor*

hot-spot factor ↦*power-peaking factor*

hot standby[1] *roc* • standby of a reactor unit after shutdown, in which the reactor temperature is kept close to its operational temperature ↦ hot standby[2]; reactor unit *D Warmzustand bei Nullast F réacteur en état d'attente à chaud Pl reaktor wyłączony w stanie gorącym Sv varmavställd reaktor*

hot standby[2] *roc* • state of a reactor

unit just before start-up, in which the reactor temperature is close to its operational temperature ↦ hot standby[1]; reactor unit *D - F état d'arrêt à chaud Pl stan gorący przy zerowej mocy Sv varmberedskap*

hot testing *nf* • (at fuel reprocessing:) testing of a method, a process, an apparatus or an instrument under normal working conditions and at expected activity levels ↦ hot; fuel reprocessing; cold testing *D heiße Prüfung; aktive Prüfung F essai en actif Pl testowanie wysokoaktywne Sv högaktiv provning*

house load operation *roc* • operation mode when turbine sets are supplying energy only for the reactor unit's own needs ↦ reactor unit *D Turbinenbetrieb zur Deckung des Eigenbedarfs F îlotage Pl praca na potrzeby własne Sv husturbindrift*

HTGR ↦*high-temperature gas-cooled reactor*

hydraulic diameter *th* • (in fluid mechanics:) characteristic length representing an equivalent diameter for a non-circular duct, defined as, $D_h = 4A/P_w$, where D_h is the *h.d.*, A is the duct cross-sectional area and P_w is its total wetted perimeter; for a circular tube, the *h.d.* is equal to the pipe inner diameter; for an infinite square lattice of fuel rods with diameter d and pitch p in a fuel assembly, the *h.d.* is obtained as $D_h = d\left[\frac{4}{\pi}\left(\frac{p}{d}\right)^2 - 1\right]$, whereas for the triangular lattice it is given as $D_h = d\left[\frac{2\sqrt{3}}{\pi}\left(\frac{p}{d}\right)^2 - 1\right]$ ↦ Reynolds number *D hydraulische Durchmesser F diamètre hydraulique Pl średnica hydrauliczna Sv hydraulisk diameter*

hydrodynamic instability *th* • coolant flow instability that can occur at certain combinations of the coolant mass flow rate and the reactor power ↦ coolant **D** *hydrodynamische Instabilität* **F** *instabilité hydrodynamique* **Pl** *niestabilność hydrodynamiczna* **Sv** *hydrodynamisk instabilitet*

hydrogen *mat* ● chemical element denoted H, with atomic number $Z=1$, relative atomic mass $A_r=1.00794$, density 0.0708 g/cm^3, melting point -259.34 °C, boiling point -252.87 °C, crustal average abundance 1400 mg/kg and ocean abundance 1.08×10^5 mg/L; *h.* has good neutron moderation properties with $\sigma_a/\sigma_s \sim 0.014$, where σ_a - the microscopic cross section for absorption, σ_s - the microscopic cross section for scattering ↦ element; moderate; microscopic cross section **D** *Wasserstoff* **F** *hydrogène* **Pl** *wodór* **Sv** *väte*

hydrogen embrittlement *mat* ● deterioration of ductility of some metals when exposed to hydrogen gas, of special importance for the zirconium alloys used as fuel-rod cladding in water-cooled reactors ↦ hydrogen; zirconium; irradiation embrittlement **D** *Wasserstoffversprödung* **F** *fragilisation par l'hydrogène* **Pl** *kruchość wodorowa* **Sv** *väteförsprödning*

Ii

I $\mapsto$*iodine*

ICFM $\mapsto$*in-core fuel management*

ICRP $\mapsto$*International Commission on Radiological Protection*

ideal cascade *nf* • (for isotope separation:) thought arrangement of separating stages of different sizes which yields a given separation with the lowest possible mass flow rate $\mapsto$ isotope separation **D** *ideale Kaskade* **F** *cascade idéale* **Pl** *kaskada idealna* **Sv** *ideal kaskad*

ideal gas *th* • hypothetical gas that satisfies: (i) the ideal gas law, (ii) Avogadro's law and which (iii) has a constant specific heat; the *ideal gas law* (also known as *Clapeyron's equation of state*, formulated by him in 1834) is given as $pV = nRT$, where p is the pressure of the gas, V is the volume of the gas, n is the amount of substance of gas in moles, R is the universal gas constant and T is the absolute temperature of the gas $\mapsto$ equation of state; universal gas constant; absolute temperature **D** *ideales Gas* **F** *gaz parfait* **Pl** *gaz idealny* **Sv** *ideal gas*

ideal gas law $\mapsto$*ideal gas*

IGSCC $\mapsto$*intergranular stress corrosion cracking*

immaterial particle $\mapsto$*particle*

importance function *rph* • function which gives the asymptotic mean number of neutrons which, in a critical system, originate from one neutron moving at a given location with a given speed; the *i.f.* is proportional to the adjoint flux $\mapsto$ relative importance; adjoint flux **D** *Einflußfunktion* **F** *fonction importance* **Pl** *funkcja cenności (neutronów)* **Sv** *importansfunktion*

incident *rs* • any unintended event, including operating errors, equipment failures, initiating events, accident precursors, near misses or other mishaps, or unauthorised act, malicious or non-malicious, the consequences or potential consequences of which are not negligible from the point of view of protection or safety; incident is often used in the same meaning as *minor accident* $\mapsto$ accident; event; INES **D** *Vorfall* **F** *incident* **Pl** *incydent* **Sv** *incident*

incineration *wst* • (for waste treatment:) transformation of wastes through heating and oxidation in order to reduce their volume $\mapsto$ waste treatment **D** *Veraschung* **F** *incinération* **Pl** *spopielanie* **Sv** *förbränning*

in-core fuel management *roc* • (abbreviated *ICFM:*) calculation in advance of reactivity and other parameters for different core configurations under one or more operating periods in order to get the best fuel utilization in the reactor $\mapsto$ operating period; reactivity **D** *Brennstoffeinsatzplanung* **F** *gestion du combustible en pile* **Pl** *gospodarka paliwem w rdzeniu* **Sv** *härdbränsleplanering*

indirect-cycle reactor *rty* • reactor in which heat is transferred from the primary coolant to the secondary coolant before a conversion of the heat to the required energy form occurs ↦ direct-cycle reactor; primary coolant; secondary coolant **D** *Indirektkreisreaktor* **F** *réacteur à cycle indirect* **Pl** *reaktor z obiegiem pośrednim* **Sv** *reaktor med indirekt cykel; indirektcykelreaktor*

indirectly ionizing particle *rd* • uncharged particle which can release directly ionized particles or cause transmutation, e.g., a neutron or a photon ↦ directly ionizing particle; transmutation **D** *indirekt ionisierendes Teilschen* **F** *particule indirectement ionisante* **Pl** *cząstka pośrednio jonizująca* **Sv** *indirekt joniserande partikel*

individual dosimeter *rdp* • dosimeter that is used to determine the individual dose equivalent to which a person is exposed ↦ dosimeter; individual dose equivalent; penetrating individual dose equivalent; superficial individual dose equivalent **D** *Individualdosimeter* **F** *dosimètre individuel* **Pl** *dawkomierz osobisty* **Sv** *persondosimeter*

induced radioactivity *rdy* • (for a nuclide:) radioactivity that is created through irradiation, usually by neutrons; the *i.r.* is of concern for several elements present in structural materials of nuclear reactors, such as the major components of steels (iron, chromium and nickel) and zirconium, as a component of zircaloy alloys used for cladding; additional elements of concern that may be present as minor alloying additions or as impurities in various steels are manganese, cobalt, copper, zinc, molybdenum, tantalum and tungsten ↦ nuclide; irradiation; neutron **D** *induzierte Radioaktivität* **F** *radioactivité induite* **Pl** *promieniotwórczość sztuczna* **Sv** *inducerad radioaktivitet*

inelastic scattering *nap* • scattering during which the total kinetic energy of a particle changes; the *i.s.* can occur in different ways: (i) during a radiative inelastic scattering, a part of the kinetic energy is used to excite the target nucleus, which returns to non-excited state by emitting a photon; (ii) during thermal inelastic scattering, energy exchange takes place between, e.g., a thermal neutron and a molecule or a crystal grid, which causes change of the energy state of the molecule or of the grid ↦ scattering; thermal neutron **D** *unelastische Streuung* **F** *diffusion inélastique* **Pl** *rozpraszanie niesprężyste* **Sv** *inelastisk spridning*

inerted containment ↦*containment*[1]

INES *rs* • (acronym for *I*nternational *N*uclear *E*vent *S*cale:) designed for promptly communicating to the public in consistent terms the safety significance of events at nuclear installations; it contains the following eight levels: level 0 - deviation; level 1 - anomaly; level 2 - incident; level 3 - serious incident; level 4 - accident without significant off-site risk; level 5 - accident with off-site risk; level 6 - serious accident; level 7 - major accident ↦ event; nuclear installation; incident; accident **D** *Internationale Bewertungsskala für nukleare Ereignisse* **F** *échelle internationale des événements nucléaires* **Pl** *skala INES* **Sv** *INES skala*

infinite multiplication constant ↦*infinite multiplication factor*

infinite multiplication factor *rph* • (denoted k_∞:) multiplication factor for a hypothetical system of infinite size; since there is no loss of neutrons by leakage, the *i.m.f.* is equal to the ratio of the total number of neutrons produced in one generation to the total number of neutrons absorbed in the preceding generation in an infinitely large system ↦ multiplication factor; four-factor formula **D** *unendlicher Multiplikationsfaktor; unendlicher Vermehrungsfaktor* **F** *facteur de multiplication infini* **Pl** *współczynnik mnożenia neutronów w ośrodku nieskończonym* **Sv** *oändlig multiplikationskonstant*

inherently stable reactor *rs* • nuclear reactor that is inherently able to compensate for possible positive reactivity changes through built-in, negative reactivity feedbacks, before any process that is destructive for the reactor can take place ↦ reactivity feedback; reactor stability **D** *inhärenter stabiler Reaktor* **F** *réacteur à stabilité intrinsèque* **Pl** *reaktor z naturalną stabilnością* **Sv** *inherent stabil reaktor*

inherent safety *rs* • engineered safety which results from laws of nature and which means that risk of a certain dangerous event is eliminated ↦ engineered safety; passive safety **D** - **F** - **Pl** *bezpieczeństwo naturalne* **Sv** *inherent säkerhet*

inhour equation *rph* • (in reactor theory, derived from "inverse hour":) equation which provides a relationship between reactivity and a time constant of a reactor, given as:

$$\rho_0 = \frac{sl}{sl+1} + \frac{1}{sl+1} \sum_{i=0}^{6} \frac{s\beta_i}{s+\lambda_i} \equiv \rho(s)$$

where ρ_0 - reactivity inserted to the reactor, β_i - delayed neutron fraction from ith precursor group, l - mean lifetime of prompt neutron in reactor, λ_i - decay constant of ith precursor group and s - Laplace transformation parameter ↦ reactivity; reactor time constant; Laplace transformation **D** *Inhour-Gleichung* **F** *équation de l'inhour* **Pl** *równanie inhour; równanie stałej czasowej* **Sv** *tidkonstanteqvation*

initial conversion ratio *rph* • conversion ratio in a nuclear reactor before any significant fuel burnup has occurred ↦ conversion ratio; burnup **D** *anfängliches Konversionsverhältnis* **F** *rapport de conversion initial* **Pl** *początkowy współczynnik konwersji* **Sv** *initialt konversionsförhållande; ursprungligt konversionsförhållande*

initial core *rph* • reactor core that contains its first fuel charge ↦ fuel charge **D** *Erstkern* **F** *coeur initial* **Pl** *rdzeń początkowy* **Sv** *initialhärd*

initiating event *rs* • (also called *initiator*:) first event in a sequence of events which can lead to core damage, such as, e.g., loss of off-site power ↦ core damage; loss of off-site power **D** *auslösendes Ereignis* **F** *événement initiateur; initiateur* **Pl** *zdarzenie inicjujące* **Sv** *inledande händelse*

initiator ↦ *initiating event*

installed capacity *roc* • (for a reactor unit:) rated output of turbogenerator(s) of the reactor unit ↦ reactor unit; rated output; turbogenerator; capacity factor **D** *installierte Leistung* **F** *puissance installée* **Pl** *moc zainstalowana* **Sv** *installerad effekt*

integral reactivity *rph* • (for a certain control rod position:) reactivity change which is achieved when the control rod is removed from the core ↦ control rod; differential reactivity **D** *Integralreaktivität* **F** *réactivité intégrale* **Pl** *reaktywność całkowa* **Sv** *integral reaktivitet*

integral reactor *rty* • reactor with a heat exchanger located between the primary and secondary coolant circuit, which is placed inside the reactor vessel ↦ primary-coolant circuit; secondary-coolant circuit; reactor vessel **D** *Reaktor mit integrierten Wärmetauschern* **F** *réacteur à échangeur intégré* **Pl** *reaktor z* *wbudowanym wymiennikiem ciepła* **Sv** *reaktor med inbyggd värmeväxlare*

interfacial slip ratio *th* • (for two-phase flow in a channel:) ratio of the cross-section mean gas velocity to the cross-section mean liquid velocity in the channel ↦ two-phase flow **D - F - Pl** *poślizg międzyfazowy* **Sv** *slipfaktor*

intergranular stress corrosion cracking *mat* • (abbreviated *IGSCC*:) stress-corrosion cracking along grain borders in a material ↦ stress-corrosion cracking; transgranular stress-corrosion cracking **D** *interkristalline Spannungsrißkorrosion* **F** *fissuration intergranulaire sous contrainte* **Pl** *pękanie wskutek korozji naprężeniowej międzykrystalicznej* **Sv** *interkristallin spänningskorrosionssprickning*

interim storage *wst* • engineered storage of nuclear waste when waiting for processing, transport or ultimate storage ↦ engineered storage; ultimate storage; nuclear waste **D** *Zwischenlagerung* **F** *dépôt*[2] **Pl** *składowanie tymczasowe* **Sv** *mellanlagring; avfallslagring*

interlock circuit *roc* • logic circuit in a control system that blocks certain actions before determined conditions are satisfied ↦ interlock limit; control rod **D** *Sperrkreis* **F** *circuit de verrouillage* **Pl** *obwód blokady* **Sv** *förreglingskrets; förbudskrets*

interlock limit *roc* • limit value of an operational parameter which, when reached, blocks certain operations, such as additional withdrawal of control rods ↦ interlock circuit; control rod **D** *Sperrgrenze* **F** *valeur limite de blockage; valeur de blockage* **Pl** *wartość graniczna blokady* **Sv** *förreglingsgräns*

intermediate cooling circuit *th* • (in nuclear power plant:) cooling circuit which transports heat from the primary circuit to a working fluid or to the final circuit; the *i.c.c.* prevents direct contact between the (possibly) radioactive coolant and the working fluid or the final cooling water (for example the sea water) ↦ reactor unit **D** *Zwischenkühlkreislauf* **F** *circuit de refroidissement intermédiaire* **Pl** *pośredni obieg chłodzenia* **Sv** *mellankylkrets*

intermediate-level waste *wst* • radioactive waste with high enough activity that it must be shielded, but not cooled ↦ radioactive waste; activity **D** *mittelaktiver Abfall* **F** *déchet de moyenne activité* **Pl** *odpad średnioaktywny* **Sv** *medelaktivt avfall*

intermediate neutron *rph* • neutron with kinetic energy between the fast and thermal neutron energy regions; in reactor physics, it customarily corresponds to the energy region between 0.1 MeV and 1 eV ↦ eV; thermal neutron energy; fast neutron **D** *mittelschnelles Neutron* **F** *neutron intermédiaire* **Pl** *neutron pośredni* **Sv** *intermediär neutron*

intermediate range monitor *roc* • power-range monitor for power range between the source range and the operating range ↦ power-range monitor; source range; operating range **D** *Zwischenbereichmonitor* **F** *moniteur du domaine intermédiaire* **Pl** *monitor zakresu mocy pośredniej* **Sv** *mellaneffektkanal*

intermediate reactor *rty* • (also called *intermediate spectrum reactor*:) nuclear reactor in which fissions are mainly caused by intermediate neutrons ↦ intermediate neutron; fission **D** *mittelschneller Reaktor* **F** *réacteur à neutrons intermediaires; réacteur à spectre intermediaire* **Pl** *reaktor na neutronach pośrednich; reaktor pośredni* **Sv** *intermediär reaktor*

intermediate spectrum reactor ↦*intermediate reactor*

internal consumption *roc* • electrical energy that is used for operation of the auxiliary machinery and other equipment in a reactor unit or another electrical installation ↦ reactor unit *D Eigenverbrauch F consommation propre Pl zużycie na potrzeby własne Sv egenförbrukning*

internal contamination *rdp* • undesirable radioactive substances absorbed into a human body ↦ radioactive *D interne Kontamination F contamination interne Pl skażenie wewnętrzne Sv intern kontamination*

internal energy *th* • energy stored in matter including its thermal energy and the binding energy of molecules, electrons in atoms and nucleons in atomic nuclei ↦ thermal energy; binding energy *D innere Energie F énergie interne Pl energia wewnętrzna Sv inre energi*

internal event *rs* • initiating event caused by a failure inside of an installation ↦ event; external event; initiating event *D internes Ereignis F événement interne Pl zdarzenie wewnętrzne Sv inre händelse*

internal irradiation *rdp* • irradiation of a biological object with radiation emitted by a radioactive substance that is inside the object ↦ irradiation; external irradiation *D innere Bestrahlung F irradiation interne Pl napromienianie wewnętrzne Sv intern bestrålning*

internal peaking factor *th* • (for a reactor with vertical fuel assemblies:) ratio of the maximum to mean power value in a horizontal cross section of a fuel assembly or fuel element ↦ axial peaking factor; power-peaking factor; radial peaking factor *D interner Formfaktor F facteur (de forme) interne Pl wewnętrzny współczynnik rozkładu mocy Sv intern formfaktor*

internal primary recirculation

pump *rcs* • pump which, partly or entirely, is inserted in the reactor vessel and which is designed to circulate the coolant ↦ reactor vessel; coolant *D interne Hauptkühlmittelpumpe F pompe interne de circulation Pl wewnętrzna główna pompa cyrkulacyjna Sv intern huvudcirkulationspump; instickspump*

internal source term ↦radioactive *source term*

International Commission on Radiological Protection *rdp* • (abbreviated *ICRP:*) free-standing, but cooperating with IAEA and WHO, an international organ with a goal to establish recommendations for radiation protection purposes ↦ radiation protection *D ICRP F ICRP Pl Międzynarodowa Komisja Ochrony Radiologicznej Sv Internationella strålskyddskommissionen*

international nuclear event scale ↦*INES*

invariant imbedding *rph* • (in transport theory:) mathematical method with which integral parameters, such as, e.g., reflection coefficients, can be calculated for a certain system without previous calculation of the detailed neutron flux density distribution within the system ↦ transport theory; neutron flux density *D invariante Einbettung F immersion invariante Pl - Sv invariant inbäddning*

inventory change *sfg* • (in safeguards of nuclear materials:) increase or decrease of nuclear material in the material balance area ↦ safeguards (of nuclear materials); nuclear material *D Bestandsänderung F variation de stock Pl zmiana zasobu Sv inventarieförändring*

iodine *mat* • chemical element denoted I, with atomic number $Z=53$, relative atomic mass $A_r=126.90447$, density 4.93 g/cm^3, melting point 113.7 °C, boiling point 184.4 °C, crustal average abundance 0.45 mg/kg and ocean abundance 0.06 mg/L; ^{133}I and ^{134}I are important fission products from the point of view of their high activity; ^{135}I is a fission product decaying into ^{135}Xe and thus contributing to the xenon poisoning ↦ element; fission product; xenon poisoning **D** *Iod* **F** *iode* **Pl** *jod* **Sv** *jod*

ion *nap* • atom or a group of atoms with an electric net charge; sometimes the definition includes a free electron as well ↦ electron **D** *Ion* **F** *ion* **Pl** *jon* **Sv** *jon*

ionization chamber *mt* • gas-filled detector of radiation whose operation is based on collection of all the charges created by direct ionization within the gas through the application of an electric field ↦ radiation detector; fission chamber **D** *Ionisationskammer* **F** *chambre d'ionisation* **Pl** *komora jonizacyjna* **Sv** *jonkammare; jonisationskammare*

ionize *nap* • completely remove an electron from an atom ↦ atom; electron; ion **D** *ionisieren* **F** *ioniser* **Pl** *jonizować* **Sv** *jonisera*

ionizing radiation *rd* • radiation that consists of directly or indirectly ionized particles; interactions of directly ionizing radiation produce ionization and excitation of the medium; indirectly ionizing radiation cannot ionize atoms but can cause interactions whose charged products, known as secondary radiation, are directly ionizing; as far as regulations and radiation protection are concerned, visible and ultraviolet light is usually excluded ↦ directly ionizing particle; indirectly ionizing particle **D** *ionisierende Strahlung* **F** *rayonnement ionisant* **Pl** *promieniowanie jonizujące* **Sv** *joniserande strålning*

iron *mat* • chemical element denoted

Fe, with atomic number $Z=26$, relative atomic mass $A_r=55.845$, density 7.875 g/cm^3, melting point 1538 °C, boiling point 2861 °C, crustal average abundance 5.63×10^4 mg/kg and ocean abundance 0.002 mg/L; *i.* is frequently present in construction and shielding materials in nuclear power plants ↦ element; radiation shielding **D** *Eisen* **F** *fer* **Pl** *żelazo* **Sv** *järn*

iron ore concrete *mat* • high-density concrete that contains crushed iron ore to improve its shielding properties ↦ shield **D** *Eisenerzbeton* **F** *béton au minerai de fer* **Pl** *beton z rudą żelaza* **Sv** *järnmalmsbetong*

irradiation ↦*exposure*^{1}

irradiation assisted stress corrosion *mat* • (abbreviated *IASC:*) stress corrosion for which ionizing radiation is one of the contributing factors ↦ stress corrosion; ionizing radiation **D** *strahlungsinduzierte Spannungskorrosion* **F** *-* **Pl** *korozja naprężeniowa indukowana napromienieniem* **Sv** *strålningsinducerad spänningskorrosion*

irradiation channel *rcs* • channel penetrating through a shield inside a nuclear reactor, in which a sample can be irradiated ↦ shield; irradiation **D** *Bestrahlungskanal* **F** *canal d'irradiation* **Pl** *kanał do napromieniania* **Sv** *bestrålningskanal*

irradiation embrittlement *mat* • deterioration of material ductility due to irradiation ↦ irradiation **D** *Bestrahlungsversprödung* **F** *fragilisation par irradiation* **Pl** *kruchość wskutek napromienienia* **Sv** *bestrålningsförsprödning*

irradiation reactor *rty* • nuclear reactor that is mainly used as a radiation source to irradiate material samples or for medical applications; example of the *i.r.* is the material testing reactor and the production reactor ↦ production reactor; materials testing reactor; irradiation **D** *Bestrahlungsreaktor* **F** *réacteur d'irradiation* **Pl** *reaktor do napromieniania* **Sv** *bestrålningsreaktor*

irradiation rig *rcs* • device placed in a reactor, containing material samples for experimental irradiation, as well as instrumentation to measure (and sometimes to control) the conditions under which the irradiation takes place ↦ irradiation **D** *Bestrahlungsvorrichtung* **F** *installation d'irradiation* **Pl** *urządzenie do napromieniania* **Sv** *bestrålningsrigg*

irreversible process *th* • thermodynamic process which, due to losses, cannot be reversed ↦ reversible process **D** *Irreversibler Prozess* **F** *processus irréversible* **Pl** *proces nieodwracalny* **Sv** *irreversibel process*

isobar *th* • curve that connects points with the same value of pressure ↦ Brayton cycle **D** *Isobar* **F** *isobare* **Pl** *izobara* **Sv** *isobar*

isodose *rdp* • surface or curve that connects points with the same value of the absorbed dose ↦ absorbed dose **D** *Isodose* **F** *isodose* **Pl** *izodoza* **Sv** *isodos*

isolation *rs* • closure of isolation valves, usually automatic, to limit consequences of a possible core damage ↦ isolation valve **D** *Isolierung* **F** *isolement* **Pl** *izolacja (reaktora); oddzielenie reaktora* **Sv** *reaktorisolering*

isolation valve *rcs* • valve in a pipeline that penetrates the containment wall; typically there is an outer and an inner isolation valve in each duct, in direct proximity of the containment wall; closure of isolation valves leads to an isolation of the reactor ↦ containment; isolation **D** *Verschlußventil* **F** *vanne d'isolement* **Pl** *zawór odcinający* **Sv** *skalventil*

isotope *bph* • one of several nuclides with the same atomic number but different mass numbers ↦ nuclide; atomic number; mass number **D** *Isotop* **F** *isotope* **Pl** *izotop* **Sv** *isotop*

isotope separation *nf* • separation of one or several isotopes from the other isotopes that are present in the element ↦ enrichment[1]; isotope; natural abundance; abundance ratio **D** *Isotopentrennung* **F** *séparation isotopique* **Pl** *separacja izotopów* **Sv** *isotopseparation; separation*[1]

isotopic abundance *bph* • (in a mixture of isotopes of an element:) ratio of the number of atoms of a certain isotope to the total number of atoms ↦ isotope; natural abundance; abundance ratio **D** *Isotopenhäufigkeit* **F** *teneur isotopique* **Pl** *zawartość izotopowa* **Sv** *isotophalt*

ITER *rty* • (acronym for *I*nternational *T*hermonuclear *E*nergy *R*eactor, or Latin for "the way":) fusion reactor under construction at Cadarache, in the south of France; *I.* is designed to produce 500 MW of fusion power from D-T reactions for extended periods of time (5 to 7 min.) and to produce 10 times more power than is needed to maintain the plasma, achieving a *Q*-factor of 10 ↦ fusion **D** *ITER* **F** *ITER* **Pl** *ITER* **Sv** *ITER*

Jj

JET *rty* • (acronym for *J*oint *E*ropean *T*orus:) fusion reactor of tokamak-type, in operation at Cullham Laboratory since 1983, which produced a power of 1.7 MW from fusion reactions ↦ fusion **D** *JET* **F** *JET* **Pl** *JET* **Sv** *JET*

jet pump *rcs* • device to circulate coolant through the reactor core using external pumps to create driving stream in a downcomer of a reactor vessel ↦ main circulation pump **D** *Strahlpumpe* **F** *éjecteur* **Pl** *pompa strumieniowa; strumienica* **Sv** *strålpump*

joule *bph* • the SI unit of energy or work, equal to the work done by a force of one newton when its point of application moves one meter in the direction of action of the force; 1 joule $= 1\ J = 1\ N\ m = 1\ kg\ m^2\ s^{-2}$ ↦ energy; newton **D** *Joule* **F** *joule* **Pl** *dżul* **Sv** *joule*

Kk

$K^1 \mapsto potassium$

$K^2 \mapsto kelvin$

KBS-3 technology $\mapsto bedrock\ deposi$-
tory

kelvin *th* • (denoted K:) base SI unit
of thermodynamic temperature, defined by taking the fixed numerical
value of the Boltzmann constant when
expressed in the unit joule/kelvin
(J/K) $\mapsto$ Boltzmann constant; Celsius
(temperature) scale; Fahrenheit (tempera-
ture) scale **D** *kelvin* **F** *kelvin* **Pl** *kelvin* **Sv**
kelvin

kerma *rdp* • (acronym for *K*inetic
*E*nergy of *R*adiation absorbed per
unit *MA*ss, in a given material:) total
initial kinetic energy of charged particles which, per material mass, has
been released by indirectly ionizing
particles; *k.* is a deterministic quantity given as,

$$K \equiv \lim_{\Delta m \to 0} \frac{\Delta \bar{E}_{tr}}{\Delta m},$$

where $\bar{E}_{tr}$ is the expected or stochas-
tic average sum of the initial kinetic
energies of all the charged ionizing
particles released by interaction of indirectly ionizing particles in matter of
mass m; in many practical situations
the *k.* is a good approximation of the
absorbed dose $\mapsto$ indirectly ionizing particle; absorbed dose **D** *Kerma* **F** *kerma* **Pl**
kerma **Sv** *kerma*

keV $\mapsto electron\ volt$

key measurement point *sfg* • (in
safeguards of nuclear materials, abbreviated *KMP:*) place in a nuclear
installation where nuclear material
exists in such form that the material
flow can be easily determined $\mapsto$ safeguards; nuclear material; nuclear installation
D *Schlüsselmesspunkt* **F** *point de mesure
principal* **Pl** *kluczowy punkt pomiarowy* **Sv**
mätningspunkt

**Kirchhoff's law of thermal radi-
ation** *th* • radiation law saying that
for an opaque body (that is a body
that does not transmit any radiation)
the absorptivity of the body is equal
to its emissivity for every wavelength
of light $\mapsto$ thermal radiation **D** *Kirch-
hoffsches Strahlungsgesetz* **F** *loi du rayonnement de Kirchhoff* **Pl** *prawo promieniowania Kirchhoffa* **Sv** *Kirchhoffs strålningslag*

Kirchhoff's transformation *th* •
integral transformation used in a non-
linear heat conduction equation, when
the thermal conductivity is a function
of temperature $\lambda = \lambda(T)$; for a non-
stationary heat conduction problem,
the non-linearity is significantly weak-
ened, whereas for stationary heat conduction problems it is completely removed; a non-linear heat conduction
equation,

$$\varrho c_p \frac{\partial T}{\partial t} = \nabla\left[\lambda\left(T\right) \nabla T\right],$$

can be transformed into

$$\frac{\partial \vartheta}{\partial t} = a(\vartheta)\nabla^2 \vartheta,$$

where

$$\vartheta = T_0 + \frac{1}{\lambda_0} \int_{T_0}^{T} \lambda(T')dT',$$

T_0 is a reference temperature and $\lambda_0 = \lambda(T_0)$; the *K.t.* can be used to solve the heat conduction equation in oxide fuels $\mapsto$ thermal conductivity; heat conduction **D** *Kirchhoff-Transformation* **F** *transformation de Kirchhoff* **Pl** *transformacja Kirchhoffa* **Sv** *Kirchhofftransform*

KMP $\mapsto$*key measurement point*

knocked-on atom *mat* • atom which is displaced from its initial position through a collision with a particle $\mapsto$ atom; particle **D** *Rückstoßatom* **F** *atome percuté* **Pl** *atom wybity* **Sv** *stötatom*

Ll

labyrinth *rdp* • shields arranged in such a way that an opening exists but the radiation cannot reach directly from the source to the surroundings ↦ shield   **D** *Labyrinth* **F** *labyrinthe* **Pl** *labirynt* **Sv** *labyrint*

laminar flow *th* • viscous flow structure characterized by smooth motion in laminae or layers, with no macroscopic mixing of adjacent fluid layers; in a case of flow through a pipe, the nature of the flow (laminar or turbulent) is determined by the value of the Reynolds number, Re; pipe flow is laminar when Re $\leq$ 2300 and it may be turbulent for larger values ↦ Reynolds number; turbulent flow   **D** *laminare Strömung* **F** *écoulement laminaire* **Pl** *przepływ laminarny* **Sv** *laminär strömning*

landfill ↦*controlled tipping*

landfilling ↦*controlled tipping*

Laplace operator ↦*Laplacian operator*

Laplace transform *mth* • image function $F(s)$ of a complex variable $s = x + iy$ associated with an original function $f(t)$ of a real variable t through the Laplace transformation ↦ Laplace transformation   **D** *Laplace-Transformirte* **F** *transformée de Laplace* **Pl** *transformata Laplace'a* **Sv** *Laplacetransform*

Laplace transformation *mth* • association of a unique result or image function $F(s)$ of the complex variable

$s = x + iy$ with every single-valued object or original function $f(t)$ of a real variable t such that the following improper integral exists,

$$F(s) \equiv L[f(t)] \equiv \int_0^\infty f(t)e^{-st}dt$$

$F(s)$ is called the (one-sided) Laplace transform of $f(t)$; *L.t.* is mainly used to solve differential equations; original functions and their corresponding image functions are commonly provided in tables ↦ Laplace transform   **D** *Laplace-Transformation* **F** *transformation de Laplace* **Pl** *przekształcenie Laplace'a* **Sv** *Laplacetransformationen*

Laplacian operator *mth* • (also called *del square*, *nabla square* or *Laplace operator*:) differential operator ∇^2 equivalent to double usage of the nabla operator; in three-dimensional space with coordinates x, y, z, the operator is given as $\nabla^2 \equiv \frac{\partial^2}{\partial x^2} + \frac{\partial^2}{\partial y^2} + \frac{\partial^2}{\partial z^2}$; the *L.o.* occurs in differential equations describing the diffusion and heat conduction processes ↦ nabla; diffusion equation; heat conduction   **D** *Laplace-Operator* **F** *opérateur laplacien* **Pl** *laplasjan* **Sv** *Laplaceoperatorn*

laser process *nf* • isotope separation process based on differences in the light absorption spectra for isotopes of certain elements, for example hydrogen, uranium and plutonium; the *l.p.* selectively excites the target isotope, which later is separated using physical or chemical methods $\mapsto$ isotope separation; enrichment[1] *D Laserverfahren zur Isotopentrennung F procédé de séparation isotopique par laser; procédé laser Pl proces laserowy (separacji izotopów) Sv laserprocess*

latent heat $\mapsto$*heat of vaporization*

laws of thermodynamics *th* • four laws of thermodynamics including: (i) zeroth law on systems in mutual equilibrium, (ii) first law on energy conservation, (iii) second law on the increase of entropy, (iv) third law on the entropy approaching a constant value as the temperature approaches absolute zero $\mapsto$ energy *D Hauptsätze der Thermodynamik F principes de la thermodynamique Pl zasady termodynamiki Sv termodynamikens huvudsatser*

lead *mat* • chemical element denoted Pb, with atomic number $Z=82$, relative atomic mass $A_r=207.2$, density 11.342 g/cm^3, melting point 327.46 °C, boiling point 1749 °C, crustal average abundance 14 mg/kg and ocean abundance 3×10^{-5} mg/L; *l.* is an important shielding material and, in the liquid form, is used as the coolant in the lead-cooled fast reactor $\mapsto$ element; shield; coolant; lead-cooled fast reactor *D Blei F plombe Pl ołów Sv bly*

lead brick *rdp* • brick-shaped element of a shield made of lead $\mapsto$ shield; lead *D Bleiziegel; Bleistein F brique de plombe Pl cegła ołowiowa Sv blytegel; blysten*

lead-cooled fast reactor *rty* • (abbreviated *LFR*:) generation IV reactor concept with the fast neutron spectrum in which the liquid lead or lead-bismuth alloy is used as the coolant $\mapsto$ generation IV reactor; lead *D schneller bleigekühlter Reaktor F réacteur rapide refroidi au plomb Pl reaktor prędki*

chłodzony ołowiem Sv blykyld snabb reaktor

leakage[1] *rph* • (within reactor theory:) net loss of neutrons from a region due to crossing the region boundary $\mapsto$ shield; leakage[2] *D Durchlaßstrahlung; Ausfluß F fuite de neutrons Pl ucieczka (neutronów) Sv läckning*[1]

leakage[2] *rdp* • passage of radiation through a shield, in particular through cracks or other imperfections in the shield $\mapsto$ shield; leakage[1] *D Leckstrahlung F fuite Pl przeciek (promieniowania) Sv strålläckning; läckning*[2]

LET $\mapsto$*linear energy transfer*

lethargy *rph* • natural logarithm of a ratio of a reference neutron energy E_0 (usually chosen to be the maximum energy that neutrons can achieve in the system) and the actual neutron energy E: $u \equiv \ln(E_0/E)$ $\mapsto$ neutron energy distribution *D Lethargie F léthargie Pl letarg Sv letargi*

LEU $\mapsto$*low enriched uranium*

LFR $\mapsto$*lead-cooled fast reactor*

Li $\mapsto$*lithium*

light-water reactor *rty* • (abbreviated *LWR*:) reactor moderated with ordinary water $\mapsto$ heavy-water reactor; boiling water reactor *D leichtwassermoderierter Reaktor F réacteur à eau ordinaire Pl reaktor lekkowodny Sv lättvattenreaktor*

limiting condition *roc* • condition imposed on certain parameters, which leads to a limitation of the reactor power; examples of the *l.c.* include: the heat flux on the clad surface, the DNB ratio and the clad temperature $\mapsto$ DNB ratio *D begränzende Bedingung F condition limitative Pl warunek ograniczający Sv begränsande villkor*

linear energy transfer *nap* • (abbreviated *LET*, for a certain kind of charged particles passing through a substance:) averaged energy loss per unit particle path length that particles experience due to collisions, where the energy transfer rate is below a certain value; the severity and permanence of the biological damage created by ionizing radiation is directly related to the *l.e.t.* and large values of the *l.e.t.* tend to result in greater biological damage ↦ particle; absorbed dose; dose equivalent **D** *lineares Energieübertragungsvermögen* **F** *Transfert Linéique d'Énergie* **Pl** *liniowe przenoszenie energii* **Sv** *längdenergiöverföring; LET; linear energiöverföring*

linear extrapolation distance *rph* ● (also called *extrapolation length*, usually denoted *d*:) distance from a medium surface to a point where the tangent to the asymptotic neutron flux density distribution at the medium surface, calculated with the one-group model, crosses the zero line; for a plane surface at a vacuum boundary, the *l.e.d.* is approximately given by $d = 0.7104\lambda_{tr}$, where λ_{tr} is the transport mean free path of the medium ↦ asymptotic neutron flux density; one-group model; transport mean free path **D** *linearer Extrapolationsabstand* **F** *distance d'extrapolation linéaire* **Pl** *liniowa długość ekstrapolowana* **Sv** *linjär extrapolationslängd*

linear heat rate ↦*linear power density*

linear power ↦*linear power density*

linear power density *th* ● (usually denoted q', also called *linear power* or *linear heat rate*:) thermal power generated per unit length of a fuel rod or a fuel element; for a fuel rod with fuel pellet radius r_F and volumetric heat source in fuel q''', the *l.p.d.* is given as: $q' = \pi r_F^2 q'''$ ↦ fuel rod; fuel element **D** *lineare Stableistungsdichte* **F** *puissance linéique* **Pl** *liniowa gęs-*

tość mocy; moc liniowa **Sv** *längdeffekt; lineareffekt; längdvärmebelastning*

liquid metal cooled reactor *rty* ● nuclear reactor in which a liquid metal coolant, e.g., liquid sodium or liquid lead, is used ↦ lead-cooled fast reactor; sodium-cooled fast reactor; coolant **D** *flüssigmetallgekühlter Reaktor* **F** *réacteur refroidi par métal liquide* **Pl** *reaktor chłodzony ciekłym metalem* **Sv** *smältmetallkyld reaktor*

lithium *mat* ● chemical element denoted Li, with atomic number $Z=3$, relative atomic mass $A_r=6.941$, density 0.534 g/cm^3, melting point 180.5 °C, boiling point 1342 °C, crustal average abundance 20 mg/kg and ocean abundance 0.18 mg/L; a rare (7.4%) isotope of lithium (^{6}Li) can be used in thermal reactors to produce tritium (t), which is needed for the most promising (deuterium-tritium) fusion reaction on Earth, according to: $n_{th} + {}^6\text{Li} \to \alpha + t$ (exothermic reaction); tritium can also be produced in neutron collisions with the common (92.6%) isotope of lithium ^{7}Li: $n + {}^7\text{Li} \to t + \alpha + n$ (endothermic reaction) ↦ element; tritium; nuclear fusion **D** *Lithium* **F** *lithium* **Pl** *lit* **Sv** *litium*

load following *roc* ● adjusting of a reactor power level according to variations of the electric power demand ↦ reactor control **D** *Lastfolge* **F** *suivi de charge* **Pl** *obciążenie nadążające* **Sv** *lastföljning*

loading pattern *roc* ● method to distribute fuel assemblies in a reactor core during refuelling and shuffling, e.g., the scatter loading ↦ refuelling; shuffling; scatter loading **D** *Beladeplan* **F** *plan de chargement* **Pl** *plan załadunku* **Sv** *laddmönster; laddningsmönster*

LOCA ↦*loss-of-coolant accident*

local criticality *rph* ● criticality

achieved in a fraction of a nuclear reactor core ↦ global criticality **D** *lokale Kritikalität* **F** *criticité locale* **Pl** *krytyczność lokalna* **Sv** *lokal kriticitet*

local power range monitor *mt* • (abbreviated *LPRM:*) fixed neutron detectors, usually more than one hundred fission ion chambers operating in the current mode, distributed throughout the core and measuring the neutron flux in range from 2 to 150% of the full power ↦ collectron; travelling in-core probe system **D** - **F** - **Pl** *lokalny monitor zakresu mocy* **Sv** *LPRM-system*

log-normal distribution *mth* • probability distribution of a random variable $\eta = e^\xi$, where ξ has a normal distribution; the probability density function of the *l.-n.d.* is given as, $f_\eta(x; \mu, \sigma) = \frac{1}{\sigma x \sqrt{2\pi}} e^{\frac{-(\ln x - \mu)^2}{2\sigma^2}}$ for $x > 0$ and $f_\eta(x; \mu, \sigma) = 0$ for $x \leq 0$; here, μ is the mean and σ is the standard deviation of the random variable η; the *l.-n.d.* describes reasonably well the drop size distribution in two-phase dispersed flows; it is applied in the risk and safety analysis of nuclear systems to represent the flood damage magnitude and earthquake magnitude and frequency distributions ↦ normal distribution; probability density function; two-phase flow; probabilistic safety analysis **D** *logarithmische Normalverteilung* **F** *loi log-normale* **Pl** *rozkład logarytmiczny normalny* **Sv** *lognormalfördelning*

long-lived nuclide *wst* • (related to waste management:) radioactive nuclide with a radioactive half-life longer than about 30 years ↦ waste management; nuclide; radioactive half-life **D** *langlebiges Radionuklid* **F** *nucléide à vie long* **Pl** *długożyciowy nuklid* **Sv** *långlivad nuklid*

long-lived waste *wst* • radioactive

waste which contains long-lived nuclides ↦ radioactive waste; long-lived nuclide **D** *langlebiger radioaktiver Abfall* **F** *déchet à vie long* **Pl** *długożyciowy odpad radioaktywny* **Sv** *långlivat avfall*

loop reactor *rty* • liquid metal cooled fast spectrum reactor in which the intermediate heat exchanger (and possibly other parts of the primary loop) is connected to the reactor core with free pipelines ↦ pool reactor **D** *Loop-Reaktor* **F** *réacteur à boucles* **Pl** *reaktor pętlowy; reaktor z obiegiem zewnętrznym* **Sv** *externkretsreaktor*

loss-of-coolant accident *rs* • (abbreviated *LOCA:*) core accident resulting in a loss of reactor coolant at a high rate; the *l.o.c.a.* can occur due to, e.g., a pipe rupture of the reactor coolant system ↦ core accident; recirculation; emergency core cooling; coolant make-up; primary system **D** *Kühlmittelverlustunfall* **F** *accident de perte de réfrigérant primaire* **Pl** *awaria ucieczki chłodziwa* **Sv** *kylmedelshaveri*

loss of off-site power *roc* • malfunction that can be caused by a short circuit or a wire break and which results in an inability of the external power grid to accept electrical power from the plant or to supply electrical power to the plant ↦ station blackout **D** *Netzabwurf* **F** *perte du réseau électrique; perte d'alimentation du réseau; perte du réseau* **Pl** *utrata zasilania elektrycznego; utrata zasilania* **Sv** *nätbortfall*

low enriched uranium *nf* • (abbreviated *LEU:*) uranium in which the fraction of ^{235}U is less than 5% ↦ highly enriched uranium **D** *schwachangereichertes Uran* **F** *uranium faiblement enrichi* **Pl** *uran niskowzbogacony* **Sv** *låganrikat uran*

lower drywell *rcs* • (for boiling water reactor:) part of a drywell in a containment located under the reactor vessel ↦ boiling water reactor; containment; drywell; reactor vessel; pedestal *D unterer Primärkreisraum* *F enceinte sèche inférieure* *Pl dolna komora sucha* *Sv undre primärutrymme*

lower plenum *rcs* • part of a reactor vessel located under the core ↦ reactor core; reactor vessel *D* - *F* - *Pl dolna komora mieszania* *Sv undre plenum*

low-level waste *wst* • radioactive waste with so low activity that it doesn't require shielding or cooling ↦ radioactive waste *D schwachaktiver Abfall* *F déchet radioactif de faible activité; déchet de faible activité* *Pl odpad niskoaktywny* *Sv lågaktivt avfall*

LPRM ↦ *local power range monitor*

LWR ↦ *light-water reactor*

Lyapunov stability *mth* • system is Lyapunov-stable when a small perturbation of the system is moving it to a state which is close to the initial state before the perturbation; the *L.s.* theory is applicable to the reactor stability analysis ↦ reactor stability *D Ljapunow-Stabilität* *F stabilité de Lyapunov* *Pl stabilność Lapunowa* *Sv Ljapunov-stabilitet*

Mm

macroscopic *rph* • (adjective used in reactor theory:) which indicates the spatial distribution of a reactor parameter (for example, the neutron flux density) with smoothing of local effects due to the reactor cell averaging ↦ reactor cell[1]; microscopic **D** *makroskopisch* **F** *macroscopique* **Pl** *makroskopowy* **Sv** *makroskopisk*

macroscopic cross section *xr* • (denoted Σ:) cross section per unit volume, calculated as $\Sigma = N\sigma$, where N - atom density of a given material and σ - microscopic cross section of that material ↦ cross section; atom density; microscopic cross section **D** *makroskopischer Wirkungsquerschnitt* **F** *section efficace macroscopique* **Pl** *makroskopowy przekrój czynny* **Sv** *makroskopiskt tvärsnitt*

macroscopic transport cross section ↦ *transport cross section*

magnesium *mat* • chemical element denoted Mg, with atomic number $Z=12$, relative atomic mass $A_r=24.3050$, density 0.74 g/cm^3, melting point 650 °C, boiling point 1090 °C, crustal average abundance 2.33×10^4 mg/kg and ocean abundance 1290 mg/L ↦ magnox; element **D** *Magnesium* **F** *magnésium* **Pl** *magnez* **Sv** *magnesium*

magnox *mat* • (acronym for *mag*nesium *non-ox*idising:) magnesium alloy with low content of aluminum, used as a clad material in magnox reactors; *m.* has low cross section for neutron capture, but its disadvantage is that it allows for low operating temperatures (below 360 °C) and reacts with water, preventing long-term storage of spent fuel under water in spent fuel pools ↦ magnesium; aluminum; magnox reactor **D** *Magnox* **F** *magnox* **Pl** *magnoks* **Sv** *magnox*

magnox reactor *rty* • first-generation of a gas-cooled power reactor in UK, called after the magnesium-aluminium alloy, which was used as the cladding material; the *m.r.* used metallic natural uranium as fuel, CO_2 as the coolant and graphite as the moderator; 26 such reactors at 11 sites were in operation in the UK between 1956 and 2015 and the last one (Wylfa Unit 1) was closed on December 30, 2015 ↦ magnox; magnesium; power reactor **D** *Magnoxreaktor* **F** *réacteur magnox* **Pl** *reaktor magnoksowy* **Sv** *magnoxreaktor*

main circulation pump *rcs* • pump used in the primary-coolant circuit to circulate coolant through the reactor core ↦ primary-coolant system; primary-coolant circuit; primary system; reactor core **D** *Primärwasserpumpe* **F** *pompe primaire de circulation* **Pl** *główna pompa cyrkulacyjna* **Sv** *huvudcirkulationspump; HC-pump*

major actinide ↦ *minor actinide*

manganese *mat* • chemical element denoted Mn, with atomic number $Z=25$, relative atomic mass $A_r=54.938049$, density 7.3 g/cm³, melting point 1246 °C, boiling point 2061 °C, crustal average abundance 950 mg/kg and ocean abundance 2×10^{-4} mg/L ↦ element **D** *Mangan* **F** *manganèse* **Pl** *mangan* **Sv** *mangan*

manipulator *rdp* • tool which transforms movements of hands, fingers or feet into movements that perform certain operations at another place, in particular in hot cells, thereby avoiding heavy finger or hand doses ↦ hot cell **D** *Manipulator* **F** *télémanipulateur* **Pl** *manipulator* **Sv** *manipulator*

Markoczy correlation *th* • correlation used to calculate the Nusselt number for convective heat transfer in rod bundles as $Nu = C \cdot Nu_{DB}$, where Nu_{DB} is the Nusselt number found from the Dittus-Boelter correlation and $C = 1 + 0.91Re^{-0.1}Pr^{0.4}\left(1 - 2e^{-B}\right)$, where Re is the Reynolds number, Pr is the Prandtl number and $B = D_h/d$, with D_h - bundle hydraulic diameter and d - rod diameter ↦ Dittus-Boelter correlation; Weisman correlation; Reynolds number; Prandtl number; Nusselt number; hydraulic diameter **D** *Markoczy Gleichung* **F** *corrélation de Markoczy* **Pl** *korelacja Markoczyego* **Sv** *Markoczys samband*

mass coefficient of reactivity *rph* • reactivity coefficient with respect to the mass of a certain material at a certain location in a reactor ↦ reactivity coefficient **D** *Massenkoeffizient der Reaktivität* **F** *coefficient massique de réactivité* **Pl** *masowy współczynnik reaktywności* **Sv** *reaktivitetens masskoefficient*

mass defect *bph* • amount by which the rest mass of an atomic nucleus differs from the sum of the rest masses

of nucleons ↦ atomic nucleus; nucleon **D** *Massendefekt* **F** *défaut de masse* **Pl** *defekt masy* **Sv** *massdefekt*

mass flow rate *th* • physical quantity representing the mass flowing through a certain area per unit time, expressed in kg/s in the SI units ↦ mass flux **D** - **F** *débits-masse* **Pl** *wydatek masowy* **Sv** *massflöde*

mass flux *th* • physical quantity representing the mass flowing through a unit area in a unit time, expressed in kg/m²s in the SI units ↦ mass flow rate **D** - **F** *débit-masse surfacique* **Pl** *gęstość strumienia masy; prędkość masowa* **Sv** *massflödestäthet*

mass number *bph* • (denoted A:) total number of nucleons in the atomic nucleus ↦ nucleon; atomic nucleus **D** *Massenzahl* **F** *nombre de masse* **Pl** *liczba masowa* **Sv** *masstal; nukleontal*

mass separator *nf* • device for a complete or partial separation of isotopes, for example through the electromagnetic isotope separation or the gaseous diffusion process ↦ gaseous diffusion process; isotope separation; enrichment **D** *Massentrenner* **F** *séparateur de masse* **Pl** *separator masy* **Sv** *masseparator*

material balance area *sfg* • (abbreviated MBA:) control area consisting of a nuclear installation or a part of it, where the amount of nuclear material transported into the area or out of the area can be determined and the total amount of the nuclear material can be determined according to specified procedures ↦ nuclear installation; nuclear material; material balance period **D** *Materialbilanzzone* **F** *zone de bilan matière* **Pl** *strefa bilansu materiałowego* **Sv** *materialbalansområde*

material balance period *sfg* • (abbreviated MBE:) time between two consecutive physical controls of a nuclear material ↦ material balance area; nuclear material **D** *Materialbilanzperiode* **F** *période du bilan matière* **Pl** *okres bilansu materiałowego* **Sv** *materialbalansperiod*

material buckling *rph* • (denoted B_m^2:) parameter that is a measure of the neutron multiplication properties for a medium and depends on the materials that are present in the medium and their mutual placement ↦ geometric buckling; neutron multiplication **D** *materielle Flußdichtewölbung* **F** *laplacien matière* **Pl** *parametr materiałowy reaktora* **Sv** *materialbuktighet; materiell buktighet*

material fatigue *mat* • material damage due to persistent load variations ↦ thermal fatigue **D** *Materialermüdung* **F** *fatigue (matériau)* **Pl** *zmęczenie materiału* **Sv** *materialutmattning*

material particle ↦*particle*

materials testing reactor *rty* • reactor that is used for testing the behaviour of materials and reactor components under strong radiation ↦ radiation damage **D** *Materialprüfreaktor* **F** *réacteur d'essais de matériaux* **Pl** *reaktor do testowania materiałów* **Sv** *materialprovningreaktor*

material unaccounted for *sfg* • (abbreviated *MUF*, for safeguards of nuclear materials:) difference between the nuclear material accounted for and the actual amount of nuclear material ↦ safeguards **D** *nicht nachgewiesenes Material* **F** *différence d'inventaire* **Pl** *nierozliczony materiał (jądrowy)* **Sv** *oredovisad kärnämnesmängd*

maximal power *roc* • highest power achieved in a certain period of time, determined either as an instantaneous value or as a mean value during a certain time, for example one hour ↦ power control **D** *Maximalleistung* **F**

puissance maximale **Pl** *moc maksymalna* **Sv** *maximieffekt*

maximum credible accident *rs* • (abbreviated *MCA:*) old term for the design-basis accident ↦ design-basis accident **D** *größter anzunehmender Unfall* **F** *accident maximal prévisible* **Pl** *maksymalna wiarygodna awaria* **Sv** *antaget maximalt haveri*

maximum excess reactivity *rph* • the largest value of the excess reactivity during the lifetime of a reactor core or during a certain operating period ↦ built-in reactivity; excess reactivity **D** *maximale Überschußreaktivität* **F** *excédent maximal de réactivité* **Pl** *maksymalna reaktywność nadmierna* **Sv** *maximal överskottsreaktivitet*

maximum permissible limiting value *roc* • limiting value for a certain parameter which must not be exceeded during operation of the reactor unit ↦ reactor unit; operating limiting value **D** *höchstzulässiger Grenzwert* **F** *valeur limite maximale admissible* **Pl** *maksymalna wartość dopuszczalna* **Sv** *högsta tillåtna gränsvärde; HTG*

Maxwell's distribution law *bph* • statistical kinetic energy distribution of atoms or molecules in a gas at temperature T given as:

$$N(E) = \frac{2\pi N}{(\pi kT)^{3/2}} E^{1/2} \exp\left(-\frac{E}{kT}\right)$$

where E - particle energy, $N(E)$ - density of particles per unit energy, k - Boltzmann's constant ↦ neutron energy distribution; Watt's distribution law; Boltzmann constant **D** *Maxwellverteilung* **F** *loi de distribution de Maxwell* **Pl** *rozkład Maxwella* **Sv** *Maxwell-fördelning*

MBA ↦*material balance area*

MBP ↦*material balance period*

MCA ↦*maximum credible accident*

mean free path *nap* • (denoted λ:)

average path covered by a particle in a material before a collision occurs; the *m.f.p.* can refer to all types of collisions or to a particular type such as capture, scattering or ionization; for any reaction x with the macroscopic cross section Σ_x, the *m.f.p.* is $\lambda = 1/\Sigma_x$; in particular, the total mean free path is $\lambda = 1/\Sigma_t$, where Σ_t is the total macroscopic cross section ↦ transport mean free path; capture; scattering; ionize; mean life; macroscopic cross section **D** *mittlere freie Weglänge* **F** *libre parcours moyen* **Pl** *średnia droga swobodna* **Sv** *fri medelväglängd; medelfriväg*

mean life *rdy* • average time during which a particle or a particle system exists in a particular state; the *m.l.* can also refer to the average time when a particle exists in a specific region (e.g., a neutron in a reactor core) ↦ mean free path **D** *mittlere Lebensdauer* **F** *vie moyenne* **Pl** *średni czas życia* **Sv** *medellivslängd*

median lethal dose *rdp* • (usually denoted LD$_{50}$:) absorbed dose that within a given time, leads to death of 50% of individuals in a large population of a given art ↦ absorbed dose **D** *mittlere Letaldosis* **F** *dose létal 50 pourcent* **Pl** *dawka połowicznej śmiertelności* **Sv** *medianletaldos; LD 50*

median lethal time *rdp* • time that passes before 50% of individuals in a large population of a given art die after absorbing a specific dose ↦ absorbed dose; median lethal dose **D** *Medianletalzeit* **F** *temps létal 50 pourcent* **Pl** *czas połowicznej śmiertelności* **Sv** *medianletaltid*

medical radiology ↦ *radiology*

medium enriched uranium *nf* • (abbreviated *MEU:*) uranium in which the fraction of $^{235}_{92}U$ is higher than 5% and less than 20%; the *m.e.u.* cannot be used for weapon produc-

tion without an additional isotope separation ↦ isotope separation; low enriched uranium; highly enriched uranium **D** *Uran mit mittlerer Anreicherung* **F** *uranium moyennement enrichi* **Pl** *uran średniowzbogacony* **Sv** *medelanrikat uran*

medium-level waste ↦ *intermediate-level waste*

member for automatic control *roc* • control element which is a part of the control system ↦ control element **D** *Element der automatischen Steuerung* **F** *élément de réglage automatique* **Pl** *element regulacji automatycznej* **Sv** *reglerelement*

MEU ↦ *medium enriched uranium*

MeV ↦ *electron volt*

Mg ↦ *magnesium*

microscopic *rph* • (within reactor theory:) which describes the local variations of a reactor quantity (such as, e.g., neutron flux) within a reactor cell ↦ macroscopic; reactor cell[1] **D** *mikroskopisch* **F** *microscopique* **Pl** *mikroskopowy* **Sv** *mikroskopisk*

microscopic cross section *xr* • (denoted σ, expressed in barns:) cross section per single atom nucleus, characterizing the probability of neutron-nuclear reaction for the nucleus ↦ cross section; barn; macroscopic cross section **D** *mikroskopischer Wirkungsquerschnitt* **F** *section efficace microscopique* **Pl** *mikroskopowy przekrój czynny* **Sv** *mikroskopiskt tvärsnitt*

migration *wst* • (about radioactive nuclides:) change of the position of the radioactive nuclide in the bedrock ↦ nuclide **D** *Transfer* **F** *migration* **Pl** *migracja* **Sv** *migration*

migration area *rph* • (usually denoted M^2_{th}:) sum of the slowing-down area from the source energy to the thermal energy (or "age to thermal", τ_{th}) and a square of the diffusion length for the thermal neutrons (L^2_{th}):

$M_{th}^2 = L_{th}^2 + \tau_{th}$ ↦ slowing-down area; diffusion length; age; thermal neutron energy **D** *Wanderfläche* **F** *aire de migration* **Pl** *powierzchnia migracji* **Sv** *migrationsarea*
migration length *rph* • square root of the migration area; the *m.l.* for thermal neutrons in four common moderators at 293 K are as follows: water - 0.059 m, heavy water - 1.01 m, beryllium - 0.233 m and graphite (reactor grade) - 0.575 m ↦ migration area; thermal neutrons; moderator **D** *Wanderlänge* **F** *longueur de migration* **Pl** *długość migracji* **Sv** *migrationslängd*

mill *gnt* • one-thousandth of a US dollar; the *m.* is an internationally-used unit to calculate, e.g., the energy generating cost ↦ energy **D** *Mill* **F** *mill* **Pl** *mill* **Sv** *mill*

minimum critical infinite cylinder *rph* • (for a given fissile material:) infinite cylinder with the minimum diameter which can be made critical through mixing the material with another arbitrary material without any additional limitations as far as material composition, location in space and presence of moderators or reflectors are concerned ↦ minimum critical infinite slab; minimum critical mass; minimum critical volume; fissile material; critical; moderator; reflector **D** *minimaler kritischer Zylinderdurchmesser* **F** *diamètre critique minimale d'un cylindre infini* **Pl** *najmniejszy krytyczny nieskonczony cylinder* **Sv** *minsta kritiska oändliga cylinder*

minimum critical infinite slab *rph* • (for a given fissile material:) infinite slab with the minimum thickness which can be made critical through mixing the material with another arbitrary material without any additional limitations as far as material composition, location in space and presence of moderators or reflectors

are concerned ↦ minimum critical infinite cylinder; minimum critical mass; minimum critical volume; fissile material; critical; moderator; reflector **D** *minimale kritische Schichthöhe* **F** *épaisseur critique minimale de plaque infinie* **Pl** *najcieńsza krytyczna nieskonczona płyta* **Sv** *minsta kritiska oändliga skiva*

minimum critical mass *rph* • (for a given fissile material:) minimum mass of the material which can be made critical without any limitations as far as material composition, location in space and presence of moderators or reflectors are concerned ↦ minimum critical infinite cylinder; minimum critical infinite slab; minimum critical volume; fissile material; critical; moderator; reflector **D** *minimale kritische Masse* **F** *masse critique minimale* **Pl** *minimalna masa krytyczna* **Sv** *minsta kritiska massa*

minimum critical volume *rph* • (for a given fissile material:) minimum volume of the material which can be made critical through mixing this material with another arbitrary material without any additional limitations as far as material composition, location in space and presence of moderators or reflectors are concerned ↦ minimum critical infinite cylinder; minimum critical infinite slab; minimum critical mass; fissile material; critical; moderator; reflector **D** *minimales kritisches Volumen* **F** *volume critique minimal* **Pl** *najmniejsza objętość krytyczna* **Sv** *minsta kritiska volym*

minor accident ↦*incident*

minor actinide *bph* • actinide element in used nuclear fuel other than uranium and plutonium (referred to as the *major actinides*), e.g., neptunium, americium, curium ↦ actinide; uranium; plutonium; neptunium; americium; curium **D** *minore Actinoide* **F** *actinide mineur* **Pl** *aktynowiec* *mniejszościowy* **Sv** *mindre aktinid; mindre aktinoid*

mist flow *th* • two-phase flow pattern consisting of the continuous gas phase and the dispersed liquid phase ↦ two-phase flow **D** - **F** *écoulement à brouillard* **Pl** *przepływ mgłowy* **Sv** *droppflöde*

mitigative system *rs* • safety system designed to limit a release of radioactive material and thus to mitigate consequences of a core accident ↦ core accident; radioactive material **D** *abschwächendes System* **F** *systeme modérateur* **Pl** *system zmniejszania skutków (awarii rdzenia)* **Sv** *utsläppsbegränsande system; konsekvenslindrande system*

mixed convection ↦*convection*

mixed oxide fuel *nf* • (abbreviated *MOX-fuel:*) nuclear fuel consisting of a mixture of uranium and plutonium oxides ↦ uranium; plutonium **D** *Mischoxidbrennstoff; MOX-Brennstoff* **F** *combustible MOX* **Pl** *paliwo MOX; paliwo mieszanych tlenków* **Sv** *blandoxidbränsle; MOX-bränsle*

mixer-settler *nf* • device used for liquid-liquid extraction; the *m.-s.* consists of a mixing chamber, in which a water solution is mixed with the organic phase, and a settling chamber, in which the phases are separated from each other thanks to their different densities ↦ extraction cycle **D** *Mischabsetzer* **F** *mélangeur-décanteur* **Pl** *mieszalnik-osadnik* **Sv** *mixer-settler*

mixing-cup quality ↦*steam quality*

Mn ↦*manganese*

moderate *rph* • (about neutrons:) slowdown through collisions with nuclei in a material with low neutron absorption ↦ neutron absorption **D** *moderieren* **F** *modérer* **Pl** *moderować; spowalniać* **Sv** *moderera*

moderating ratio *rph* • (for a moderator:) ratio of a neutron slowing-down power to a thermal macroscopic absorption cross section ↦ slowing-down power; neutron absorption cross section; moderator **D** *Bremsverhältnis* **F** *efficasité d'un modérateur* **Pl** *współczynnik spowalniania* **Sv** *modereringskvot*

moderator *rph* • material that contains light atoms possessing a large scattering cross section and a low-absorption cross section, used to moderate neutrons ↦ scattering cross section; moderate **D** *Moderator* **F** *modérateur* **Pl** *moderator* **Sv** *moderator*

moderator control *roc* • reactor control through change of properties, location or amount of the moderator ↦ reactor control[2]; moderator; moderator dumping **D** *Moderatortrimmung* **F** *commande par le modérateur* **Pl** *sterowanie (reaktora) moderatorem* **Sv** *moderatorstyrning*

moderator dumping *roc* • reactor trip of a heavy-water reactor by tapping the moderator ↦ reactor trip; heavy-water reactor; reactor control[2]; moderator **D** *Moderatorschnellablaß* **F** *évacuation du modérateur; vidange du modérateur* **Pl** *zrzut moderatora* **Sv** *moderatoravtappning*

moderator tank *rcs* • cylindrical tank located inside the reactor pressure vessel, which separates the coolant flowing through the core from the coolant entering the pressure vessel and flowing down in the downcomer towards the lower plenum; for PWRs, the term *core barrel* is used ↦ reactor vessel; downcomer; lower plenum **D** *Moderatorbehälter* **F** *cuve du modérateur* **Pl** *kosz rdzenia; zbiornik moderatora* **Sv** *moderatortank*

mode switch *roc* • component used to select an operation form belonging to the reactor protective system, for example reactor start-up or reactor normal operation $\mapsto$ protective system; reactor start-up; normal operation **D** Betriebsformwähler **F** sélecteur de fonctions **Pl** przełącznik trybu pracy **Sv** driftomkopplare

modified Bessel differential equation mth • ordinary differential equation:

$$\frac{d^2y}{dx^2} + \frac{1}{x}\frac{dy}{dx} - \left(\alpha^2 + \frac{n^2}{x^2}\right)y = 0$$

with a general solution of the equation given as $y(x) = C_1 I_n(\alpha x) + C_2 K_n(\alpha x)$ where C_1 and C_2 are constants, I_n is the modified Bessel function of the first kind and n-th order and K_n is the modified Bessel function of the second kind and n-th order $\mapsto$ modified Bessel functions of the first kind; modified Bessel functions of the second kind **D** modifizierte Besselsche Differentialgleichung **F** équation différentielle de Bessel modifiée **Pl** zmodyfikowane równanie różniczkowe Bessela **Sv** modifierad Bessel differentialekvationen

modified Bessel functions of the first kind mth • solutions of the modified Bessel's differential equation, expressed in terms of the Bessel functions of the first kind as follows,

$$I_n(x) = i^{-n} J_n(ix) = i^n J_n(-ix)$$

where i is the imaginary unit and $J_n(x)$ is the Bessel function of the first kind; useful approximations for small x:

$$I_0(x) = 1 + \frac{x^2}{4} + \frac{x^4}{64} + \frac{x^6}{2304} + ...$$

$$I_1(x) = \frac{x}{2} + \frac{x^3}{16} + \frac{x^5}{384} + ...$$

asymptotic expansions for large x:

$$I_0(x) = \frac{e^x}{\sqrt{2\pi x}}\left(1 + \frac{1}{8x} + ...\right)$$

$$I_1(x) = \frac{e^x}{\sqrt{2\pi x}}\left(1 - \frac{3}{8x} + ...\right)$$

$\mapsto$ Bessel functions of the first kind **D** modifizierte Bessel-Funktionen erster Gattung **F** fonctions de Bessel modifiée de première espèce **Pl** zmodyfikowane funkcje Bessela pierwszego rodzaju **Sv** modifierade besselfunktionerna av första slaget

modified Bessel functions of the second kind mth • solutions of the modified Bessel's differential equation, expressed in terms of the Bessel functions of the first and second kind as follows,

$$K_n(x) = \frac{\pi}{2}i^{n+1}\left[J_n(ix) + iY_n(ix)\right]$$

where i is the imaginary unit, $J_n(x)$ is Bessel's function of the first kind and $Y_n(x)$ is the Bessel function of the second kind; useful approximations for small x:

$$K_0(x) = -\left(\gamma + \ln\frac{x}{2}\right)I_0(x) + \frac{x^2}{4} + ...$$

$$K_1(x) = \left(\gamma + \ln\frac{x}{2}\right)I_1(x) + \frac{1}{x} - \frac{x}{4} - ...$$

with $\gamma = 0.577216$; asymptotic expansions for large x:

$$K_0(x) = \sqrt{\frac{\pi}{2x}}e^{-x}\left(1 - \frac{1}{8x} + ...\right)$$

$$K_1(x) = \sqrt{\frac{\pi}{2x}}e^{-x}\left(1 + \frac{3}{8x} + ...\right)$$

$\mapsto$ Bessel functions of the first kind; Bessel functions of the second kind **D** modifizierte Bessel-Funktionen zweiter Gattung **F** fonctions de Bessel modifiée de deuxième espèce **Pl** zmodyfikowane funkcje Bessela drugiego rodzaju **Sv** modifierade besselfunktionerna av andra slaget

moisture separator rcs • device designed to separate the liquid phase, usually as liquid droplets, from the saturated vapour phase; the m.s. is placed before the inlet to a turbine to improve the steam quality, and thus, to improve the turbine thermodynamic efficiency; in particular, a device that contains both the *m.s.* and in which the cycle steam is superheated by the live or extraction steam, is called the *moisture separator reheater* ↦ steam separator; steam quality; steam dryer **D** *Wasserabscheider* **F** *séparateur d'humidité*[1] **Pl** *separator wilgoci* **Sv** *fuktavskiljare*[1]

moisture separator reheater ↦*moisture separator*

molar mass *rph* • mass of one mole of a chemical substance expressed in the SI unit kg/mol, but due to historical reasons, frequently expressed in g/mol; since most commonly the chemical substance contains many isotopes of the same element, the *m.m.* is computed from standard atomic weights and is thus a terrestrial average and a function of the relative abundance of the isotopes on Earth ↦ mole; molecular mass; atomic mass **D** *molare Masse* **F** *masse molaire* **Pl** *masa molowa* **Sv** *molmassa*

mole *rph* • (denoted mol:) unit of measurement for amount of substance in the SI units, containing a fixed number of elementary entities (atoms or molecules) with a numerical value of Avogadro's constant, when expressed in the unit mol^{-1}, called Avogadro's number ↦ molar mass; molecular mass; Avogadro's constant; Avogadro's number **D** *Mol* **F** *mole* **Pl** *mol* **Sv** *mol*

molecular distillation *nf* • isotope separation process which is based on evaporation of a liquid mixture of isotopes in a vacuum ↦ isotope separation **D** *Molekulardestillation* **F** *distillation moléculaire* **Pl** *destylacja molekularna;*

destylacja cząsteczkowa **Sv** *molekylardestillation*

molecular mass *rph* • mass of a single neutral molecule, often expressed in the non-SI unit dalton (symbol Da or u), computed from the atomic masses of the constituent atoms; similarly, *molecular weight* is computed from the atomic weights of the constituent elements ↦ molar mass; atomic mass **D** *Molekülmasse* **F** *masse moléculaire* **Pl** *masa cząsteczkowa* **Sv** *molekylmassa*

molecular weight ↦*molecular mass*

molten salt reactor *rty* • (abbreviated *MSR:*) generation IV nuclear reactor concept whose nuclear fuel is made of molten salt with a fissile component ↦ generation IV reactor; nuclear fuel; fissile **D** *Flüssigsalzreaktor; Salzschmelzen-Reaktor* **F** *réacteur (nucléaire) à sels fondus* **Pl** *reaktor chłodzony mieszaniną stopionych soli* **Sv** *saltsmältereaktor*

molybdenum *mat* • chemical element denoted Mo, with atomic number $Z=42$, relative atomic mass $A_r=95.94$, density 10.2 g/cm^3, melting point 2623 °C, boiling point 4639 °C, crustal average abundance 1.2 mg/kg and ocean abundance 0.01 mg/L ↦ element **D** *Molybdän* **F** *molybdène* **Pl** *molibden* **Sv** *molybden*

moments method *rph* • computational method for damping of neutron and gamma radiation in which the transport equation is used to determine the spatial moment of the flux density; the *m.m.* applies particularly to an infinite, homogeneous material and has been used in the analysis of reactor shields ↦ gamma radiation; transport equation; shield **D** *Momentenmethode* **F** *méthode des moments (spatiaux)* **Pl** *metoda momentów* **Sv** *momentmetod*

monitor[1] *rdp* • measure radiation using a radiation monitor ↦ radiation; radiation monitor **D** *überwachen* **F** *contrôler; surveiller* **Pl** *kontrolować; nadzorować* **Sv** *monitera*

monitor^2 *mt* • instrument for monitoring of, e.g., a temperature or a water level ↦ radiation monitor **D** *Monitor* **F** *moniteur* **Pl** *przyrząd kontrolny* **Sv** *vakt*

Monte Carlo method *rph* • statistical method of computation through simulation of a large number of individual events; the *M.C.m.* is used, for example, to solve neutron transport problems and to determine the particle flux distributions in a medium ↦ transport equation **D** *Monte-Carlo-Verfahren* **F** *méthode de Monte-Carlo* **Pl** *metoda Monte Carlo* **Sv** *montecarlometod*

MOX-fuel ↦ *mixed oxide fuel*

MSR ↦ *molten salt reactor*

MUF ↦ *material unaccounted for*

multigroup model *rph* • model of neutron physical calculations in which neutrons are divided into several energy groups ↦ multigroup theory; neutron energy group; one-group model; two-group model **D** *Mehrgruppenmodell* **F** *modèle à plusieurs groupes* **Pl** *model wielogrupowy* **Sv** *flergruppsmodell*

multigroup theory *rph* • neutron transport theory based on the multigroup model ↦ multigroup model;

one-group theory; two-group theory **D** *Mehrgruppentheorie* **F** *théorie à plusieurs groupes* **Pl** *teoria wielogrupowa* **Sv** *flergruppsteori*

multiple scattering *rph* • scattering that results from several single scattering events ↦ single scattering **D** *Mehrfachstreuung* **F** *diffusion multiple* **Pl** *rozpraszanie wielokrotne* **Sv** *multipelspridning*

multiplication constant ↦ *multiplication factor*

multiplication factor *rph* • (denoted k:) total mean number of generated neutrons during a certain period of time (disregarding sources whose strengths do not depend on the fission rate) divided by the mean number of neutrons which during the same period of time are lost due to absorption and leakage ↦ absorption; leakage; effective multiplication constant; infinite multiplication constant **D** *Multiplikationsfaktor* **F** *facteur de multiplication* **Pl** *współczynnik mnożenia* k **Sv** *multiplikationskonstant*

multiplying *rph* • (about a system:) adjective which states that a chain reaction of neutron-induced nuclear fissions is sustained ↦ nuclear chain-reaction; nuclear fission **D** *multiplizierend* **F** *multiplicateur* **Pl** *mnożący* **Sv** *multiplicerande*

Nn

$N^1 \mapsto$ *nitrogen*

$N^2 \mapsto$ *newton*

Na $\mapsto$ *sodium*

nabla *mth* • differential operator given in three-dimensional space with coordinates x, y, z as $\nabla = \mathbf{i}\frac{\partial}{\partial x} + \mathbf{j}\frac{\partial}{\partial y} + \mathbf{k}\frac{\partial}{\partial z}$; *n.* is used in vector calculus to create three distinct differential operators: the gradient (∇), the divergence $(\nabla\cdot)$ and the curl $(\nabla\times) \mapsto$ Fick's law; diffusion coefficient **D** *Nabla* **F** *nabla* **Pl** *nabla* **Sv** *nabla*

nabla square $\mapsto$ *Laplacian operator*

narrow gap $\mapsto$ *narrow water gap*

narrow water gap *rcs* • (for boiling water reactor, also called *narrow gap:*) space between fuel assemblies in a reactor core, not intended for control rods $\mapsto$ control-rod gap; fuel assembly; boiling water reactor **D** *enger Wasserspalt; enger Spalt* **F** *-* **Pl** *wąska szczelina wodna* **Sv** *bakspalt; smalspalt*

natural abundance *bph* • isotopic abundance of an isotope in a chemical element as naturally found on Earth $\mapsto$ isotopic abundance; isotope **D** *natürliche Isotopenhäufigkeit* **F** *teneur isotopique naturelle* **Pl** *naturalny stosunek zawartości izotopów* **Sv** *naturlig isotophalt*

natural convection *th* • (also called *free convection:*) convection heat transfer, in which the fluid is in motion only due to buoyancy forces, resulting from density gradients that are caused by temperature gradients;

this type of heat transfer plays an important role in passive safety systems that are used to evacuate heat from the reactor core; the onset of *n.c.* is determined by a specific value of the Rayleigh number $\mapsto$ convection; Rayleigh number **D** *natürliche Konvektion; freie Konvektion* **F** *convection naturelle* **Pl** *konwekcja swobodna* **Sv** *egenkonvektion*

natural radioactivity *rdy* • radioactivity other than that resulting from human actions $\mapsto$ radioelement; radioisotope; induced radioactivity **D** *natürliche Radioaktivität* **F** *radioactivité naturelle* **Pl** *promieniotwórczość naturalna* **Sv** *naturlig radioaktivitet*

natural uranium *mat* • uranium as found in nature, containing 0.7198–0.7202% of ^{235}U (α-decaying, $T_{1/2}$=704 My), 99.2739–99.2752% of ^{238}U (α-decaying, $T_{1/2}$=4470 My) and a trace (0.0050–0.0059%) of ^{234}U; the *n.u.* can be used as fuel in graphite or heavy-water moderated reactors $\mapsto$ uranium; natural-uranium reactor **D** *natürliches Uran* **F** *uranium naturel* **Pl** *uran naturalny* **Sv** *naturligt uran; natururan*

natural-uranium reactor *rty* • reactor in which the natural uranium is used as fuel $\mapsto$ natural uranium; CANDU **D** *Natururanreaktor* **F** *réacteur à uranium naturel* **Pl** *reaktor na uranie naturalnym* **Sv** *natururanreaktor*

near surface repository *wst* • underground repository located near ground level ↦ repository; underground ultimate storage **D** - **F** - **Pl** składowisko przypowierzchniowe **Sv** -

negative reactivity *rph* • (also called *deficit:*) reactivity when it is negative and causes the reactor power to decrease ↦ reactivity **D** *negative Reaktivität* **F** *réactivité négative; antiréactivité* **Pl** *reaktywność negatywna* **Sv** *underskottsreaktivitet; negativreaktivitet*

Nelkin's model *rph* • model to calculate scattering kernels for thermal neutrons in water, in which approximations are made to account for vibrations and hindered rotations of molecules ↦ scattering kernel; thermal neutron **D** *Nelkin-Modell* **F** *modèle de Nelkin* **Pl** *model Nelkina* **Sv** *Nelkins modell*

neptunium *mat* • chemical element denoted Np, with atomic number $Z = 93$, relative atomic mass A_r=237.05, density 19.38 g/cm^3, melting point 639 °C, boiling point 4174 °C; *n.* is a hard, silvery, ductile, radioactive actinide metal; ^{239}Np is a radioactive isotope (β-decaying, T$_{1/2}$=2.35 d), created in nuclear reactors fuelled with uranium ↦ chemical element **D** *Neptunium* **F** *neptunium* **Pl** *neptun* **Sv** *neptunium*

net power *roc* • electrical power supplied by a reactor unit to the external electrical grid that is equal to the gross power minus the internal power consumption of the unit ↦ reactor unit; gross power; internal consumption **D** *Nettoleistung* **F** *puissance nette* **Pl** *moc netto* **Sv** *nettoeffekt*

net production *roc* • gross production reduced with the internal consumption of the production installation ↦ production reactor; gross production; internal consumption **D** *Nettoproduk-*

tion **F** *production nette* **Pl** *produkcja netto* **Sv** *nettoproduktion*

neutron *rd* • electrically neutral subatomic particle with the rest mass of 1.675×10^{-27} kg (1.008 644 904(14) u or 939.565 63(28) MeV/c^2) ↦ thermal neutrons; nucleon **D** *Neutron* **F** *neutron* **Pl** *neutron* **Sv** *neutron*

neutron absorber *nap* • body or substance with which neutrons significantly interact in such a way that they disappear as free particles without giving rise to new neutrons ↦ neutron **D** *Neutronenabsorber* **F** *absorbant de neutrons* **Pl** *absorbent neutronów* **Sv** *neutronabsorbator*

neutron absorption *nap* • nuclear process in which the incident neutron disappears; a process, in which one or more neutrons are emitted afterwards, or together with other particles, as during fission, is considered as *n.a.* as well; however, scattering is not considered as *n.a.* ↦ neutron; neutron absorber; neutron absorption cross section **D** *Neutronenabsorption* **F** *absorption des neutrons* **Pl** *absorpcja neutronów* **Sv** *neutronabsorption*

neutron absorption cross section *xr* • cross section for the neutron absorption reaction, equivalent to the difference between the total cross section and the scattering cross section ↦ cross section; neutron absorption; total cross section; scattering cross section **D** *Neutronenabsorptionsquerschnitt* **F** *section efficace d'absorption des neutrons* **Pl** *przekrój czynny na absorpcję neutronów* **Sv** *absorptionstvärsnitt*

neutron converter *mt* • device placed in a flux of slow neutrons whose function is to produce fast neutrons ↦ slow neutron; fast neutron **D** *Neutronenkonverter* **F** *convertisseur de neutrons*

Pl *konwerter strumienia neutronów* **Sv** *neutronkonverter*

neutron current density *rd* • vector whose surface-normal component is equal to the net number of neutrons passing through the surface in the positive direction, calculated per unit time and area ↦ neutron; neutron flux density **D** *Neutronenstromdichte* **F** *densité de courant de neutrons* **Pl** *gęstość prądu neutronów* **Sv** *neutronströmtäthet*

neutron cycle *rph* • neutron history expressed in terms of mean values of energy, interaction and transport, since its birth due to fission, until its absorption or leakage from the reactor ↦ neutron; neutron flux density **D** *Neutronenzyklus* **F** *cycle des neutrons* **Pl** *cykl neutronowy* **Sv** *neutroncykel*

neutron density ↦*neutron-number density*

neutron detector *mt* • device designed to measure the neutron radiation ↦ neutron; detector; radiation **D** *Neutronendetektor* **F** *détecteur de neutrons* **Pl** *detektor neutronów* **Sv** *neutrondetektor*

neutron diffusion *rph* • process in which neutrons in a substance move from regions of high neutron density to regions of low neutron density, and during which they repeatedly undergo scattering collisions ↦ neutron density; scattering **D** *Neutronendiffusion* **F** *diffusion des neutrons* **Pl** *dyfuzja neutronów* **Sv** *neutrondiffusion*

neutron diffusion coefficient ↦*diffusion coefficient*

neutron economy *rph* • management of neutrons in a nuclear reactor in such a way that they are used as desired rather than being lost due to leakage or parasitic capture ↦ leakage; parasitic capture **D** *Neutronenökonomie* **F** *économie des neutrons* **Pl** *bilans neutronów* **Sv** *neutronekonomi*

neutron energy distribution *rph* • neutron number density or neutron flux density as a function of neutron energy; the following neutron energy distribution ranges are defined: 0 - 0.025 eV: cold neutrons; 0.025 eV: thermal neutrons; 0.025 - 0.4 eV: epithermal neutrons; 0.4 - 0.6 eV: cadmium neutrons; 0.6 - 1 eV: epicadmium neutrons; 1 - 10 eV: slow neutrons; 10 - 300 eV: resonance neutrons; 300 eV - 1 MeV: intermediate neutrons; 1 - 20 MeV: fast neutrons; >20 MeV: ultrafast neutrons ↦ neutron energy group; neutron-number density; neutron flux density **D - F - Pl** *rozkład energii neutronów* **Sv** *neutronenergifördelning*

neutron energy group *rph* • one of many groups containing neutrons within an arbitrarily chosen energy interval; each group has an assigned unique characteristic energy ↦ multigroup model **D** *Neutronenenergiegruppe* **F** *groupe de neutrons (par énergie)* **Pl** *grupa energetyczna neutronów* **Sv** *neutrongrupp*

neutron flux ↦*neutron flux density*

neutron flux density *rd* • (also called *neutron flux*, denoted ϕ:) particle flux density for neutrons; an *angular flux density* $\phi(\mathbf{r}, \hat{\mathbf{\Omega}}, E, t)$ is a scalar function of seven independent variables: the neutron location vector $\mathbf{r}$, the unit vector of the direction of neutron motion $\hat{\mathbf{\Omega}}$, the neutron energy E and the time t; for most reactor calculations an *angle-integrated neutron flux* $\phi(\mathbf{r}, E, t) = \int_{4\pi} d\hat{\mathbf{\Omega}} \varphi(\mathbf{r}, \hat{\mathbf{\Omega}}, E, t)$ is used; additionally, a separable form is often assumed in which $\phi(\mathbf{r}, E, t) = \psi(\mathbf{r}, t)\varphi(E)$, where $\varphi(E)$ is the *neutron flux energy spectrum* ↦ particle flux density; neutron current density **D** *Neutronenflußdichte; Neutronenfluß* **F** *débit de fluence de neutrons; densité de flux de* *neutrons* **Pl** *gęstość strumienia neutronów* **Sv** *neutronflödestäthet; neutronflöde*

neutron flux energy spectrum
↦*neutron flux density*

neutron group condensation *rph* • reduction of the number of neutron energy groups and calculation of the corresponding group cross sections ↦ neutron energy group; group cross section **D** *Gruppenkondensation* **F** *condensation de groupes d'énergie* **Pl** *kondensacja grupy neutronów* **Sv** *gruppkondensering; neutrongruppkondensering*

neutron importance function *rph* • probability $\phi^+(\mathbf{r}, E)$ that a neutron introduced at position $\mathbf{r}$ and having energy E will ultimately result in a fission ↦ fission **D** - **F** - **Pl** *funkcja cenności neutronów* **Sv** -

neutron lifetime *rph* • (for neutrons in a certain system such as a reactor:) mean lifetime between the birth due to, e.g., fission and the disappearance due to, e.g., absorption or leakage ↦ fission; neutron absorption; leakage **D** *mittlere Neutronenlebensdauer (im Reaktor)* **F** *vie moyenne des neutrons* **Pl** *średni czas życia neutronów* **Sv** *neutronlivslängd*

neutron multiplication *rph* • process in which a neutron generates other neutrons in a medium containing a fissionable material ↦ fissionable **D** *Neutronenmultiplikation* **F** *multiplication des neutrons* **Pl** *mnożenie neutronów* **Sv** *neutronmultiplikation*

neutron number *bph* • number of neutrons in an atomic nucleus ↦ atomic nucleus **D** *Neutronenzahl* **F** *nombre de neutrons* **Pl** *liczba neutronów (w jądrze)* **Sv** *neutrontal*

neutron-number density *rd* • number of free neutrons per unit volume; partial *n.-n.d.* can be defined for neutrons with a certain energy or moving in a certain direction ↦ neutron flux

density; neutron current density **D** *Neutronenzahldichte* **F** *nombre volumique de neutrons* **Pl** *całkowita gęstość neutronów* **Sv** *neutrontäthet*

neutron physics *bph* • branch of physics that deals with properties of neutrons and their interactions with nuclei and materials ↦ neutron **D** *Neutronenphysik* **F** *physique neutronique* **Pl** *fizyka neutronów* **Sv** *neutronfysik*

neutron radiography *mt* • radiography using neutrons ↦ radiography; neutron **D** *Neutronenradiographie* **F** *neutronographie* **Pl** *radiografia neutronowa* **Sv** *neutronradiografi*

neutron source *rd* • material or device which yields or has capacity to yield neutrons ↦ neutron **D** *Neutronenquelle* **F** *source de neutrons* **Pl** *źródło neutronów* **Sv** *neutronkälla*

neutron spectroscopy *bph* • branch of spectroscopy in which neutrons are used to investigate nuclear processes through determination of cross sections for interactions of neutrons at a specific energy with nuclei ↦ spectroscopy; cross section; nucleus **D** *Neutronenspektroskopie* **F** *spectroscopie neutronique* **Pl** *spektroskopia neutronowa* **Sv** *neutronspektroskopi*

neutron temperature *rph* • (for a neutron population described with Maxwell's distribution law:) characteristic temperature for Maxwell's distribution of neutron energies ↦ Maxwell's distribution law; neutron **D** *Neutronentemperatur* **F** *température neutronique* **Pl** *temperatura neutronów* **Sv** *neutrontemperatur*

neutron transmutation doping *mat* • creation of dopant atoms in a substance, e.g., in a semiconductor, through nuclei transmutation using neutrons ↦ nucleus; transmutation **D** *Dotierung mit Neutronen*

F dopage par transmutation de neutrons *Pl domieszkowanie przez transmutację neutronami Sv neutrondopning*

neutron transport equation
$\mapsto$transport equation

neutron yield per absorption *nap*
• (denoted η:) mean of the total number of fission neutrons, including the delayed neutrons, which result from a neutron absorption in a fissionable nuclide or in nuclear fuel; the *n.y.p.a.* is a function of the neutron energy E and can be determined as $\eta(E) = \nu(E)\sigma_f(E)/\sigma_a(E)$, where $\nu(E)$ is the neutron yield per fission, $\sigma_f(E)$ is the microscopic cross section for fission and $\sigma_a(E)$ is the microscopic cross section for absorption $\mapsto$ fission neutron; fissionable; four-factor formula; neutron yield per fission; thermal-fission factor *D Neutronenausbeute je Absorption; Eta-Faktor F facteur êta Pl wydatek neutronów na jeden neutron pochłonięty; współczynnik eta Sv neutronutbyte per absorption; fissionsfaktor*

neutron yield per fission *nap* • (denoted ν:) mean number of fission neutrons released per fission, including the delayed neutrons $\mapsto$ fission neutron; fission *D Neutronenausbeute je Spaltung F facteur nu Pl wydatek neutronów na jedeno rozszczepienie Sv neutronutbyte per fission*

newton *bph* • (denoted N:) unit of force defined as the force required to accelerate a mass of one kilogram at a rate one meter per second squared; frequently used multiples of newton include $1kN = 10^3$ N, 1 MN $= 10^6$ N, 1 GN $= 10^9$ N and 1 TN $= 10^{12}$ N $\mapsto$ dyne *D Newton F newton Pl newton Sv newton*

Newton's law of cooling $\mapsto$heat transfer coefficient

Ni $\mapsto$nickel

nickel *mat* • chemical element denoted Ni, with atomic number $Z=28$, relative atomic mass $A_r=58.6934$, density 8.912 g/cm^3, melting point 1455 °C, boiling point 2913 °C, crustal average abundance 84 mg/kg and ocean abundance 5.6×10^{-4} mg/L; *n.* impurity content in the primary circuit of LWRs is controlled and limited due to its impact on the crud deposition $\mapsto$ element; crud *D Nickel F nickel Pl nikiel Sv nickel*

nile *rph* • British unit of reactivity, where 1 nile $= 0.01$ $\mapsto$ reactivity; fission; dollar; cent *D Nile F nile Pl nile Sv nile*

nitrogen *mat* • chemical element denoted N, with atomic number $Z=7$, relative atomic mass $A_r=14.00674$, density 0.807 g/cm^3, melting point -210.0 °C, boiling point -195.79 °C, crustal average abundance 19 mg/kg and ocean abundance 0.5 mg/L $\mapsto$ element *D Stickstoff F azote Pl azot Sv kväve*

nodal method *rph* • numerical method for three-dimensional reactor core analysis in which the core is divided into homogeneous parts called nodes, where each node is assumed to interact only with neighbouring nodes $\mapsto$ core analysis *D Nodalmethode F méthode nodale Pl metoda węzłowa Sv nodalmetod*

nonelastic cross section *rph* • difference between the total cross section and the cross section for elastic scattering; *n.c.s.* should not be confused with the cross section for inelastic scattering $\mapsto$ cross section; total cross section; inelastic scattering; elastic scattering *D Wirkungsquerschnitt für nichtelastische Wechselwirkung; nichtelastischer Wirkungsquerschnitt F section efficace non-élastique Pl przekrój czynny oddziaływania nieelastycznego Sv icke-elastiskt tvärsnitt*

nonleakage probability *rph* • probability that a neutron in a reactor core doesn't leak out; the term can refer to all neutrons or to certain neutron energy groups ↦ neutron; reactor; neutron energy group; leakage[1] **D** *Verbleibwahrscheinlichkeit* **F** *probabilité de non-fuite* **Pl** *prawdopodobieństwo uniknięcia ucieczki* **Sv** *icke-läckningsannolikhet*

normal distribution *mth* ● random variable X with probability density function $f_X(x; \mu, \sigma) = \frac{1}{\sigma\sqrt{2\pi}} e^{\frac{-(x-\mu)^2}{2\sigma^2}}$ for $-\infty < x < \infty$ has normal distribution, where μ is the mean and σ is the standard deviation of the random variable X ↦ log-normal distribution; probability density function **D** *Normalverteilung* **F** *loi normale* **Pl** *rozkład normalny* **Sv** *normalfördelning*

normal operation *roc* ● operation under planned conditions ↦ operating cycle; operating range **D** *Normalbetrieb* **F** *fonctionnement normal* **Pl** *praca nominalna* **Sv** *normaldrift*

notification of unusual event *rs* ● (term sometime shortened to *unusual event:*) lowest reference level for emergency action which means that events are in progress or have occurred which indicate a potential degradation of the level of safety of the plant or indicate a security threat to facility protection has been initiated; however, no releases of radioactive material requiring offsite response or monitoring are expected unless further degradation of safety systems occurs; the purpose of this classification is to assure that the first step in future response has been carried out, to bring the operations staff to a state of readiness, and to provide systematic handling of unusual event information and decision-making ↦ reference level for emergency action **D** *Auslösung eines*

Voralarms **F** *notification d'un événement inhabituel; déclenchement d'une alarme* **Pl** *zgłoszenie wyjątkowego zdarzenia* **Sv** *upplysning*

nozzle process *nf* ● isotope separation process based on the fact that, when a gas mixture expands in a certain type of nozzle, the heavier molecules have a higher concentration at the central part of the gas stream ↦ isotope separation **D** *Trenndüsenverfahren* **F** *procédé par tuyère* **Pl** *technologia dyszowa* **Sv** *dysprocess*

NSSS ↦*nuclear steam-supply system*

nuclear *bph* ● related to atomic nucleus or nuclear processes, e.g., nuclear energy, nuclear reactor; adjective "atomic" should not be used when only atomic nucleus is meant ↦ atomic nucleus; atomic **D** *Kern-; nuklear* **F** *nucléaire* **Pl** *jądrowy; nuklearny* **Sv** *kärn-; nukleär*

nuclear accident *rs* ● accident in which damages on humans or properties are due to radioactivity or toxicity resulting from activated nuclear fuel or due to radioactivity from another radiation source in a nuclear installation ↦ nuclear installation; radiation source **D** *nuklearer Unfall* **F** *accident nucléaire* **Pl** *katastrofa nuklearna; awaria jądrowa* **Sv** *nukleär olycka*

nuclear activity *gnt* ● activity that includes construction, ownership and operation of a nuclear installation as well as fuel management and disposal of nuclear waste ↦ nuclear installation; fuel management; nuclear waste **D** *kerntechnischer Umgang* **F** *activité nucléaire* **Pl** *działalność nuklearna* **Sv** *kärnteknisk verksamhet*

nuclear battery *gnt* ● device for conversion of energy directly from radioactive decay or via a thermal cycle into the electrical energy; the used radionuclide can be, for example: ^{90}Sr, ^{210}Po or ^{239}Pu ↦ radioactive decay; fuel management; nuclear waste; strontium; polonium; plutonium **D** *Isotopenbatterie* **F** *batterie nucléaire* **Pl** *bateria jądrowa* **Sv** *radionuklidbatteri*

nuclear chain-reaction *gnt* • series of nuclear reactions in which one of the components, necessary for the reactions, is reproduced and initiates similar reactions; depending on whether the average number of new reactions, which are caused directly by one reaction in a certain system, is less than unity, equal to unity, or greater than unity, the system is called subcritical, critical or supercritical, respectively ↦ subcritical; critical; supercritical **D** *nukleare Kettenreaktion* **F** *réaction nucléaire en chaîne* **Pl** *jądrowa reakcja łańcuchowa; reakcja łańcuchowa* **Sv** *nukleär kedjereaktion; kedjereaktion*

nuclear chemistry *nch* • branch of chemistry which deals with chemical aspects of nuclear science ↦ radiochemistry **D** *Kernchemie* **F** *chimie nucléaire* **Pl** *chemia jądrowa* **Sv** *kärnkemi*

nuclear criticality safety *rs* • nuclear safety which aims at prevention of an unintentional criticality ↦ criticality **D** *Kritikalitätssicherheit* **F** *sûreté-criticité* **Pl** *bezpieczeństwo krytycznościowe* **Sv** *kriticitetssäkerhet*

nuclear disintegration *rdy* • conversion of an atomic nucleus (or a compound nucleus) due to emission of particles, or due to a fission into several lighter nuclei ↦ radioactive decay **D** *Kernzerfall* **F** *désintégration nucléaire* **Pl** *rozpad jądra* **Sv** *kärnsönderfall; sönderfall*

nuclear emulsion *mt* • photographic emulsion intended for registration of traces of individual ionizing particles ↦ ionizing radiation **D** *Kernemulsion* **F** *émulsion nucléaire* **Pl** *emulsja jądrowa* **Sv** *kärnemulsion*

nuclear energy *bph* • energy released during nuclear reaction or nuclear disintegration; the term nuclear power is sometimes used as a synonym for *n. e.* when it comes to recovery of useful energy, in particular in compositions such as a nuclear power plant ↦ nuclear reaction; nuclear power; nuclear disintegration; nuclear power plant **D** *Kernenergie*[1] **F** *énergie nucléaire*[1] **Pl** *energia jądrowa* **Sv** *kärnenergi*

nuclear engineering *gnt* • branch of engineering that deals with recovery of the nuclear energy ↦ nuclear energy **D** *Kernenergetik; Kernkrafttechnik* **F** *ingénierie nucléaire* **Pl** *technika jądrowa* **Sv** *kärnenergiteknik*

nuclear fission ↦ fission

nuclear fuel *nf* • material that contains fissionable nuclides, determined to be placed in a reactor core, so that a self-sustainable chain reaction can take place ↦ fissionable; nuclear chain-reaction; reactor core **D** *Kernbrennstoff* **F** *combustible nucléaire* **Pl** *paliwo jądrowe* **Sv** *kärnbränsle*

nuclear fusion *bph* • exothermic reaction between two light atomic nuclei which leads to reaction products from which one is a nucleus with a greater mass than any of the initial nuclei ↦ nucleus **D** *Kernfusion; Kernverschmeltzung* **F** *fusion nucléaire* **Pl** *synteza jądrowa* **Sv** *fusion; kärnsammansmältning*

nuclear heating *roc* • (during startup of a cold reactor:) increase of a reactor temperature to the operational level through the development of its own power in the critical reactor ↦ reactor start-up; critical **D** *nukleare Heizung* **F** *échauffement nucléaire* **Pl** *podgrzew jądrowy* **Sv** *nukleär värmning*

nuclear heat plant *gnt* • nuclear

installation which mainly supplies thermal energy ↦ nuclear installation *D Kernkraftwerk (für Wärmeerzeugung) F centrale nucléaire (thermique) Pl ciepłownia jądrowa Sv kärnvärmeverk*

nuclear installation *gnt* • installation for extraction of the nuclear energy, manufacturing of the nuclear fuel, the fuel reprocessing or management, or disposal of the nuclear waste ↦ nuclear energy; nuclear fuel; fuel processing; nuclear waste *D kerntechnische Anlage F installation nucléaire Pl instalacja jądrowa Sv kärnteknisk anläggning; kärnenergianläggning*

nuclear island *gnt* • (in reactor unit:) nuclear reactor with the primary process systems ↦ nuclear steam-supply system; reactor unit; primary system *D nuklearer Teil des Kernkraftwerks F ilôt nucléaire Pl część jądrowa Sv reaktordel*

nuclear material *sfg* • (mainly in legal texts:) uranium, plutonium, thorium and spent fuel which has not been placed in ultimate storage ↦ spent fuel; ultimate storage *D Kernmaterial F matière nucléaire Pl materiał jądrowy Sv kärnämne*

nuclear physics *bph* • branch of physics that deals with physical aspects of nuclear science ↦ nuclear science *D Kernphysik F physique nucléaire Pl fizyka jądrowa Sv kärnfysik*

nuclear poison *rph* • material that is either created (such as, e.g., a fission product ^{135}Xe) or added (for example, boron or cadmium) to a reactor and which absorbs neutrons through a parasitic capture ↦ xenon poisoning; neutron absorption; parasitic capture *D Reaktorgift; Neutronengift F poison nucléaire Pl trucizna reaktorowa Sv reaktorgift*

nuclear power[1] *gnt* • synonym for nuclear energy that is used in case of recovery of useful energy; the term is used in compositions such as the nuclear power plant ↦ nuclear energy; nuclear power plant; nuclear power[2] *D Kernenergie[2]; Kernkraft F énergie nucléaire[2] Pl energia jądrowa Sv kärnkraft*

nuclear power[2] *roc* • power generated in the nuclear installation ↦ nuclear installation; nuclear power[1] *D nukleare Leistung F puissance nucléaire Pl moc jądrowa Sv nukleär effekt; kärneffekt*

nuclear power plant *gnt* • nuclear installation that mainly generates electricity, with one or more reactor units at the same site ↦ nuclear installation; reactor unit *D Kernkraftwerk F centrale nucléaire Pl elektrownia jądrowa; siłownia jądrowa Sv kärnkraftverk; kärnkraftstation*

nuclear power plant simulator *roc* • installation for simulation of processes in a nuclear power plant in the real time scale; the control and registration equipment corresponds to the control room in a real reactor unit and is connected to a computer system which is programmed to simulate the reactor unit behaviour under various conditions ↦ reactor simulator; reactor unit; control room *D Kernkraftwerksimulator F simulateur de centrale nucléaire Pl symulator elektrowni jądrowej Sv kärnkraftverkssimulator*

nuclear power station *gnt* • nuclear installation which mainly delivers heat ↦ nuclear installation *D Heizkraftwerk F centrale nucléaire; centrale électronucléaire Pl elektrociepłownia jądrowa Sv kärnkraftvärmeverk*

nuclear power unit ↦ reactor unit

nuclear reaction *nap* • process, such as fission, fusion, or radioactive decay, in which two nuclei, or else an atomic nucleus and a subatomic particle, collide to produce one or more nuclides that are different from the parent nuclei ↦ fission; fusion; radioactive decay **D** *Kernreaktion* **F** *réaction nucléaire* **Pl** *reakcja jądrowa* **Sv** *kärnreaktion*

nuclear reactor ↦*reactor*

nuclear safety *rs* • (for a nuclear installation:) safety that deals with protection of humans or property against harmful effects of ionizing radiation, radioactive contamination and criticality ↦ nuclear criticality safety; nuclear installation; ionizing radiation; radioactive contamination **D** *nukleare Sicherheit* **F** *sécurité nucléaire; sûreté nucléaire* **Pl** *bezpieczeństwo nuklearne* **Sv** *nukleär säkerhet*

nuclear science *gnt* • branch of science that deals with properties and reactions of atomic nuclei and with all related phenomena ↦ nuclear technology; nuclear physics; nuclear chemistry; atomic nucleus; nuclear reaction **D** *Kernwissenschaft* **F** *science nucléaire* **Pl** *nukleonika* **Sv** *kärnvetenskap*

nuclear steam-supply system *gnt* • (abbreviated *NSSS*, in a nuclear power unit:) nuclear reactor with the primary and secondary process systems that are necessary for its operation ↦ nuclear power unit; nuclear island **D** *nukleares Dampferzeugungssystem* **F** *chaudiére nucléaire* **Pl** *układ jądrowego wytwarzania pary* **Sv** *nukleärt ånggenererande system*

nuclear superheat *th* • superheat of steam in a nuclear reactor core or outside the nuclear core using the heat from the core ↦ nuclear heating **D** *nukleare Überhitzung* **F** *surchauffe nucléaire* **Pl** *przegrzew jądrowy* **Sv** *nukleär överhettning*

nuclear technology *gnt* • branch of technology that deals with technical applications of nuclear science ↦ nuclear science **D** *Kerntechnik* **F** *génie nucléaire* **Pl** *technnika jądrowa; nukleotechnika* **Sv** *kärnteknik*

nuclear waste *wst* • spent fuel and other radioactive materials which are created in a nuclear installation and that will not be further utilized ↦ radioactive waste; spent fuel **D** *nuklearer Abfall der Kerntechnik* **F** *déchets nucléaires* **Pl** *odpad jądrowy* **Sv** *kärnavfall*

nucleate boiling *th* • boiling on a heated surface whereby vapour bubbles are formed at various locations on the surface ↦ film boiling; bulk boiling **D** *Blasensieden*[1]*; Keimsieden* **F** *ébullition nucléée* **Pl** *wrzenie pęcherzykowe* **Sv** *bubbelkokning*

nucleon *bph* • proton or neutron as part of the atomic nucleus or as a free particle ↦ proton; neutron; atomic nucleus **D** *Nukleon* **F** *nucléon* **Pl** *nukleon* **Sv** *nukleon*

nuclide *bph* • kind of atom that is characterized by a composition of the nucleus and the energy level, and whose mean life is long enough to be measured ↦ atomic nucleus; mean life **D** *Nuklid* **F** *nucléide* **Pl** *nuklid* **Sv** *nuklid*

Nusselt number *th* • (denoted Nu:) non-dimensional number used in convection heat transfer analyses to express the heat transfer coefficient, defined as $Nu = hL/\lambda$, where h is the heat transfer coefficient, L is the characteristic length and λ is the thermal conductivity of fluid ↦ convection **D** *Nusselt-Zahl* **F** *nombre de Nusselt* **Pl** *liczba Nusselta* **Sv** *Nusselts tal*

Nyquist stability criterion *rs* • stability criterion based on a plot of a system transfer function G on a $Im(G)$-versus-$Re(G)$ plane, where $Im(G)$ is the imaginary part and $Re(G)$ is the real part of the system transfer function ↦ inherently stable reactor; transient **D** *Stabilitätskriterium von Nyquist* **F** *diagramme de Nyquist* **Pl** *kryterium stabilności Nyquista* **Sv** *Nyquists stabilitetskriterium*

Oo

O ↦*oxygen*

OBE ↦*operating basis earthquake*

occupational exposure *rdp* • irradiation to which certain persons are exposed when performing their professional duties ↦ irradiation; radiation work **D** *berufliche Strahlenbelastung* **F** *exposition professionnelle (aux rayonements)* **Pl** *napromienianie zawodowe* **Sv** *yrkesbestrålning*

off-gas treatment *wst* • removal of radioactive components or other impurities from gases, before they, under controlled conditions, are released to the atmosphere ↦ decontaminate **D** *Abgasbehandlung* **F** *traitement des effluents gazeux* **Pl** *obróbka gazów odlotowych* **Sv** *avgasbehandling*

off-shore siting *gnt* • siting of a nuclear installation at sea, for example on an artificial island ↦ coastal siting **D** *Verlegung im Meer* **F** *installation en mer; établissement en mer* **Pl** *lokalizacja przybrzeżna* **Sv** *havsförläggning*

off-site alert *rdp* • intermediate reference level for emergency action which is used when the operation condition or occurred event causes the preparedness level to be raised in the emergency preparedness organisation; however, there is no threat of an immediate high emission of radioactivity ↦ reference level for emergency action **D** *Katastrophenvoralarm* **F** *alerte de site* **Pl** stan podwyższonej gotowości **Sv** *höjd beredskap*

once-through charge *nf* • fuel charge which passes only once through a reactor core without the following fuel reprocessing ↦ fuel charge; fuel reprocessing **D** *einmalige Ladung* **F** *cycle à passage unique; charge à passage unique* **Pl** *wsad jednoprzejściowy* **Sv** *engångsladdning*

one-group diffusion equation ↦*diffusion equation*

one-group model *rph* • neutron-physical computational model according to which all neutrons are assumed to have the same energy ↦ neutron; two-group model; multigroup model; one-group theory **D** *Eingruppenmodell* **F** *modèle à un seul groupe* **Pl** *model jednogrupowy* **Sv** *engruppsmodell*

one-group theory *rph* • neutron transport theory based on the one-group model ↦ neutron; one-group model; two-group theory; multigroup theory **D** *Eingruppentheorie* **F** *théorie à un groupe* **Pl** *teoria jednogrupowa* **Sv** *engruppsteori*

on-line refuelling *roc* • refuelling during reactor operation, without shutting down the reactor ↦ refuelling **D** *kontinuierliche Beladung* **F** *chargement continu* **Pl** *załadunek paliwa podczas eksploatacji* **Sv** *laddning under drift*

operating basis earthquake ↦*design-basis earthquake*

operating cycle *roc* • period that begins with a reactor start-up and then includes an operating period and the closest following revision period ↦ reactor start-up; operating period; revision period **D** *Betriebszyklus* **F** *cycle de fonctionnement* **Pl** *cykl eksploatacyjny* **Sv** *driftcykel*

operating limiting value *roc* ● (for a nuclear power plant:) limit value for an operating parameter, reaching which activates automatic protection actions ↦ maximum permissible limiting value **D** *Betriebgrenzwert* **F** *limite de fonctionnement* **Pl** *graniczna wartość eksploatacyjna* **Sv** *driftgränsvärde*

operating period *roc* ● period that begins with a reactor start-up and ends with the reactor shut-down for a normal refuelling ↦ reactor start-up; refuelling **D** *Betriebsperiode* **F** *période de fonctionnement; période d'exploitation* **Pl** *okres eksploatacyjny* **Sv** *driftperiod*

operating range *roc* ● (for reactor power:) power interval within which a reactor normally operates ↦ power range **D** *Betriebsbereich* **F** *domaine de fonctionnement* **Pl** *zakres roboczy* **Sv** *driftområde*

operating rules *roc* ● conditions and guidelines for a reactor operation as decided by the regulatory authority ↦ reactor **D** *Betriebsvorschriften* **F** *spécifications techniques d'exploitation* **Pl** *reguły eksploatacji reactora* **Sv** *säkerhetstekniska föreskrifter; STF*

out-in loading *nf* ● fuel shuffling in which the fuel is successively moved towards the center of the core, where the spent fuel is finally removed from the core ↦ shuffling; spent fuel; reactor core **D** *Umsetzung zum Zentrum* **F** *cheminement* **Pl** *przeładunek (paliwa) do środka* **Sv** *bränsleinflyttning*

overhaul *roc* ● maintenance, change, or reconstruction work at a shutdown reactor unit ↦ reactor unit; shutdown **D** *Revision* **F** *révision* **Pl** *przegląd* **Sv** *revision*

overhaul period *roc* ● time from a reactor shutdown for revision and normal refuelling to the reactor start-up ↦ reactor start-up; shutdown **D** *Revisionszeitraum* **F** *période de révision* **Pl** *okres przeglądu* **Sv** *revisionsperiod*

overmoderated *rph* ● (about multiplying system:) adjective which states that the volumetric ratio of moderator to fuel is greater than the one for which a certain physical parameter in a reactor (e.g., the material buckling) has a maximum value ↦ undermoderated; well-moderated; multiplying; material buckling **D** *übermoderiert* **F** *surmodéré* **Pl** *przemoderowany* **Sv** *övermodererad*

overpack *wst* ● additional, external container for the already contained and packaged radioactive waste ↦ radioactive waste **D** *Kompaktlagerung* **F** *suremballage* **Pl** *kapsułka zewnętrzna* **Sv** *ytterbehållare*

oxide fuel ↦*ceramic fuel*

oxygen *mat* ● chemical element denoted O, with atomic number $Z=8$, relative atomic mass $A_r=15.9994$, density 1.141 g/cm^3, melting point -218.79 °C, boiling point -182.95 °C, crustal average abundance 4.61×10^5 mg/kg and ocean abundance 8.57×10^5 mg/L ↦ element **D** *Sauerstoff* **F** *oxygène* **Pl** *tlen* **Sv** *syre*

Pp

parasitic capture *nap* • capture of neutrons without fission or any other desired process occurring ↦ capture; fission; neutron **D** *parasitärer Einfang* **F** *capture parasite* **Pl** *wychwyt pasożytniczy; wychwyt nierozszczepieniowy* **Sv** *parasitisk infångning*

partial reactor trip *roc* • reactor trip when only a fraction of reactor control rods are used ↦ reactor trip; control rod **D** *partielle Schnellabschaltung* **F** *arrêt d'urgence partiel* **Pl** *awaryjne częściowe wyłączenie reaktora* **Sv** *partiellt snabbstopp*

particle *rd* • single or composite system with a well-defined and often small rest mass; particles that have the rest mass are called *material particles*, e.g., protons or neutrons, whereas particles that have no rest mass are called *immaterial particles*, e.g., photons ↦ neutron; proton; photon **D** *Teilchen* **F** *particule* **Pl** *cząstka* **Sv** *partikel*

particle fluence *rd* • number of particles that during a certain time interval fall into a small sphere around a certain point in the space, divided by the sphere's cross-section area; the *p.f.* is identical with a time integral over the particle flux density, $\Phi(t) \equiv \int_0^t \phi(t')dt'$, where $\Phi(t)$ is the *p.f.* and $\phi(t)$ is the time-dependent particle flux density ↦ particle; particle flux density **D** *Teilchenfluenz; Teilchenflu-*

ens **F** *fluence de particules; fluence* **Pl** *fluencja cząstek* **Sv** *partikelfluens*

particle fluence rate ↦*particle flux density*

particle flux density *rd* • (also called *particle fluence rate:*) number of particles that, per unit time, fall into a small sphere around a certain point in space, divided by the cross-section area of the sphere ↦ conventional flux density; neutron flux density; particle fluence **D** *Teilchenflußdichte* **F** *débit de fluence de particules; densité de flux de particules* **Pl** *gęstość strumienia cząstek* **Sv** *partikelfluensrat; partikelflödestäthet*

partition stage *nf* • (in extraction cycle:) stage in which two or more species, e.g., uranium and plutonium, are separated from each other into two different liquid phases ↦ extraction cycle **D** *Trennstufe* **F** *étage de partage* **Pl** *etap podziału* **Sv** *fördelningssteg*

passive component *rs* • component that does not require any external influence to function, such as an electric signal or a human action ↦ passive safety; passive system **D** *passive Komponente* **F** *composant passif* **Pl** *element pasywny* **Sv** *passiv komponent*

passive safety *rs* • engineered safety which is achieved by usage of passive components and systems ↦ engineered safety; inherent safety; passive component; passive system **D** *passive Sicherheit* **F** *sécu-*

119

rité passive **Pl** *bezpieczeństwo pasywne* **Sv** *passiv säkerhet*

passive system *rs* • system composed of passive components ↦ passive component **D** *passives System* **F** *systéme passif* **Pl** *system pasywny* **Sv** *passivt system*

PCI ↦*pellet-clad interaction*

pcm *rph* • (derived from *pour cent mille:*) unit of reactivity, where 1 pcm $= 10^{-5}$ ↦ reactivity; dollar; cent **D** *pcm* **F** *pcm* **Pl** *pcm* **Sv** *pcm*

peak power *roc* • electric power taken above a certain power limit to cover load peaks ↦ base power **D** *Spitzenleistung* **F** *puissance de pointe* **Pl** *moc szczytowa* **Sv** *toppeffekt*

pebble-bed reactor *rty* • reactor in which certain or all materials important for nuclear processes (e.g., nuclear fuel, fertile material, moderator) have the shape of balls, packed in a stationary bed ↦ nuclear fuel; fertile material; moderator **D** *Kugelhaufenreaktor* **F** *réacteur à lit de boulets* **Pl** *reaktor ze złożem kulowym* **Sv** *kulbäddsreaktor*

pedestal *rcs* • (for a boiling water reactor:) space beneath the reactor vessel that constitutes a part of the drywell in the reactor containment and where, between others, actuators of control rods are located ↦ boiling water reactor; drywell; control rod; lower drywell **D** *Steuerstabgrube* **F** *cavité de la cuve du réacteur* **Pl** *wnęka napędów prętów kontrolnych* **Sv** *drivdonsgrop*

pellet-clad interaction *nf* • mechanical and chemical interactions between a fuel pellet and the cladding that surrounds it, which can lead to a creation of cracks due to stress corrosion ↦ fuel pellet; cladding; stress corrosion **D** *Brennstoff-Hüllen-Wechselwirkung* **F** *interaction pastille-gaine* **Pl** *oddziaływanie pastylki z koszulką* **Sv** *växelverkan kuts-kapsling*

PENA detector ↦*collectron*

pen dosimeter *rdp* • dosimeter of the same outer form as a pen, usually consisting of an integral ion chamber combined with a miniature electroscope ↦ dosimeter; charger-reader; ion chamber **D** *Füllhalterdosimeter* **F** *stylo dosimètre* **Pl** *dawkomierz piórowy* **Sv** *pendosimeter*

penetrating individual dose equivalent *rdp* • dose equivalent in a tissue with density 1000 kg/m^3 at the depth of 1 cm inside the body; the quantity is intended for personal dosimetry and can be related to the effective dose equivalent; the *p.i.d.e.* is expressed in sieverts ↦ dose equivalent; superficial individual dose equivalent; effective dose equivalent; sievert **D** - **F** *indice d'équivalent de dose profond* **Pl** *indywidualny równoważnik dawki dla promieniowania przenikliwego* **Sv** *djup individdosekvivalent*

percentage depth dose *rdp* • absorbed dose at a certain depth in a body expressed in relation to the absorbed dose at a reference point within the irradiated region; the *p.d.d.* is usually expressed in percent; for X and gamma radiation the location of the reference point depends on the radiation energy: for low energy the reference point is located at the surface, whereas for high energy it is located at a point that receives the maximum absorbed dose ↦ absorbed dose; superficial individual dose equivalent; penetrating individual dose equivalent; X radiation; gamma radiation **D** *relative Tiefendosis* **F** *rendement en profondeur* **Pl** *procentowa dawka głębokościowa* **Sv** *relativ djupdos; djupdos*

period meter[1] *roc* • instrument to measure the doubling time of the reactor power ↦ doubling time[1] **D** *Periodenmeßgerät* **F** *périodemètre[1]* **Pl**

miernik okresu podwojenia mocy **Sv** *fördub-blingstidsmätare*

period meter² *roc* • instrument to measure the reactor time constant ↦ reactor time constant **D** *Periodenmeter* **F** *périodemètre²* **Pl** *miernik stałej czasowej reaktora* **Sv** *tidkonstantmätare*

period range ↦*time-constant range*

personal monitoring *rdp* • monitoring of a person for external irradiation, a body activity or a radioactive contamination ↦ irradiation; radioactive contamination **D** *individuelle Überwachung* **F** *surveillance du personnel* **Pl** *dawkomierz osobisty; dawkomierz indywidualny* **Sv** *personmonitering*

phantom *rdp* • volume with tissue-equivalent material matched to represent a certain biological object; the distribution of a certain radiation quantity is usually measured within this volume ↦ tissue equivalent **D** *Phantom* **F** *fantôme* **Pl** *fantom* **Sv** *fantom*

photographic dosimeter ↦*film badge*

photoluminescence dosimeter *rdp* • dosimeter whose operation principles are based on the photoluminescence of a certain material, for example, silver phosphate glass ↦ dosimeter **D** *Photolumineszenzdosimeter* **F** *dosimètre photoluminescent* **Pl** *dozymetr fotoluminescencyjny* **Sv** *fotoluminescensdosimeter*

photon *rd* • quantum of the electromagnetic radiation ↦ particle **D** *Photon* **F** *photon* **Pl** *foton* **Sv** *foton*

photoneutron *rd* • neutron released due to interaction of a photon with an atomic nucleus ↦ neutron; photon; atomic nucleus **D** *Photoneutron* **F** *photoneutron* **Pl** *fotoneutron* **Sv** *fotoneutron*

PHWR *rty* • (acronym for *P*ressurized *H*eavy *W*ater *R*eactor:) nuclear reactor in which pressurized heavy water is used as the coolant ↦

CANDU reactor; light-water reactor; pressurized water reactor **D** *PHWR* **F** *PHWR* **Pl** *PHWR* **Sv** *PHWR*

physical inventory *sfg* • (for safeguards of nuclear materials:) amount of nuclear material at a certain time as determined by measurements or estimations according to specific procedures ↦ safeguards; nuclear material **D** *realer Bestand* **F** *inventaire physique* **Pl** *fizyczny zasób (materiału jądrowego)* **Sv** *fysisk kärnämnesmängd; fysisk mängd*

pile *rty* • outdated term for a nuclear reactor ↦ nuclear reactor **D** *Atommeiler* **F** *pile atomique* **Pl** *stos atomowy* **Sv** *atommila; atomstapel*

pile oscillator ↦*reactor oscillator*

pitch *rcs* • (of reactor lattice:) distance between centers of neighbouring fuel assemblies or fuel elements ↦ reactor lattice; fuel assembly; fuel element **D** *Gitterabstand* **F** *pas du réseau* **Pl** *podziałka* **Sv** *delning²*

Planck's constant *bph* • (denoted *h*:) one of the exact fundamental physical constants equal to $h = 6.626\,070\,15 \cdot 10^{-34}$ J·s, used in a definition of the kilogram; *P.c.* expresses a relationship between energy of a photon and its frequency ↦ photon; Stefan-Boltzmann law **D** *Planck-Konstante* **F** *constante de Planck* **Pl** *stała Plancka* **Sv** *Plancks konstant*

plasma process *nf* • isotope separation process that is based on the difference in a cyclotron frequency for ions, for example ^{235}U and ^{238}U in plasma, when the plasma is subject to a strong magnetic field and electromagnetic radiation of the radio frequency in a resonance with the cyclotron frequency for ^{235}U ↦ isotope separation **D** *Plasmaverfahren zur Isotopentrennung* **F** *procédé plasma; procédé*

R.C.I. **Pl** *proces plazmowy* **Sv** *plasmaprocess*

plume *rs* • stretches of airborne radioactive substance during release from a nuclear installation ↦ radioactive substance; nuclear installation **D** *Schadstoffwolke* **F** *panache* **Pl** *pióropusz* **Sv** *plym*

plutonium *mat* • chemical element denoted Pu, with atomic number $Z=94$, relative atomic mass $A_r=239.13$, density 19.816 g/cm^3, melting point 639.4 °C, boiling point 3228 °C; the fissile nuclide ^{239}Pu (α-decaying, $T_{1/2}$=24.11 ky) is created during the operation of a nuclear power reactor whose fuel contains ^{238}U; the fertile isotope ^{240}Pu (α-decaying, $T_{1/2}$=6.564 ky) has a large capture cross section for the production of fissile isotope ^{241}Pu (β-decaying, $T_{1/2}$=14.35 y) ↦ element; fissile; fertile **D** *Plutonium* **F** *plutonium* **Pl** *pluton* **Sv** *plutonium*

plutonium credit *nf* • economic value of plutonium in the spent fuel, where the plutonium created through a conversion is considered ↦ spent fuel; fuel consumption charge; conversion **D** *Plutoniumwert* **F** *crédit plutonium* **Pl** *wartość plutonu* **Sv** *plutoniumvärde*

plutonium reactor *rty* • reactor most of whose fuel is made of plutonium ↦ plutonium **D** *Plutoniumreaktor* **F** *réacteur au plutonium* **Pl** *reaktor plutonowy* **Sv** *plutoniumreaktor*

P$_N$-approximation ↦spherical harmonics method

pneumatic system *rcs* • (in nuclear reactor or another irradiation device:) pipeline for pneumatic transport of samples to and from the irradiation position ↦ irradiation channel **D** *Rohrpostsystem* **F** *système pneumatique* **Pl** *system pneumatyczny* **Sv** *rörpostsystem*

point-kernel method *rdp* • method to calculate attenuation of gamma radiation and, in certain cases, fast neutrons, in which the radiation source is assumed to be collocated as a number of point sources; for an isotropic point source with length damping coefficient μ, flux density at a distance r is proportional to $e^{-\mu r}/\left(4\pi r^2\right)$ ↦ attenuation; gamma radiation; fast neutron; radiation source **D** *Punktkernnäherungsmethode* **F** *méthode des noyaus ponctuels* **Pl** *metoda źródeł punktowych* **Sv** *punktkärnsmetoden*

polonium *mat* • chemical element denoted Po, with atomic number $Z=84$, relative atomic mass of longest lived isotope $A_r=[209]$, density 9.169 g/cm^3 (alpha form) and 9.398 g/cm^3 (beta form), melting point 254 °C, boiling point 962 °C, and natural occurrence limited to tiny traces of ^{210}Po in uranium ores ↦ element; uranium; nuclear battery **D** *Polonium* **F** *polonium* **Pl** *polon* **Sv** *polonium*

pool boiling *th* • boiling heat transfer on a heated surface to stationary liquid in a large volume; liquid motion is only due to natural phenomena such as buoyancy forces ↦ film boiling **D** *Behältersieden* **F** *ébullition en vase* **Pl** *wrzenie basenowe; wrzenie przy konwekcji swobodnej* **Sv** -

pool reactor^1 *rty* • reactor whose fuel assemblies are submerged in a pool containing water, which serves as the moderator, the coolant and the biological shield ↦ fuel assembly; moderator; coolant; biological shield **D** *Wasserbeckenreaktor* **F** *réacteur piscine*^{1} **Pl** *reaktor basenowy*^{1} **Sv** *bassängreaktor*

pool reactor^2 *rty* • liquid metal cooled fast spectrum reactor in which the reactor, pumps, pipelines and the intermediate heat exchanger are submerged in a vessel containing coolant ↦ loop reactor **D** *Pool-Reaktor* **F** *réacteur piscine*[2] **Pl** *reaktor basenowy*[2] **Sv** *internkretsreaktor*

positron *rd* • electron with a positive charge; *p.* is an antiparticle of the negative electron ↦ particle; electron **D** *Positron* **F** *positron* **Pl** *pozytron* **Sv** *positron*

potassium *mat* • chemical element denoted K, with atomic number $Z=19$, relative atomic mass $A_r=39.0983$, density 0.89 g/cm^3, melting point 63.38 °C, boiling point 759 °C, crustal average abundance 2.09×10^4 mg/kg and ocean abundance 399 mg/L ↦ element **D** *Kalium* **F** *potassium* **Pl** *potas* **Sv** *kalium*

power coefficient of reactivity *rph* • reactivity coefficient with respect to the reactor power ↦ reactivity coefficient **D** *Leistungskoeffizient* **F** *coefficient de puissance* **Pl** *mocowy współczynnik reaktywności* **Sv** *effektkoefficient*

power control *roc* • automatic control of an output gross power of a reactor unit so that the power is kept equal to the set-point value ↦ reactor unit; gross power **D** *Leistungsregelung* **F** *régulation de puissance* **Pl** *regulacja mocy* **Sv** *effektreglering*

power density *th* • generated thermal power per unit volume ↦ specific power **D** *Leistungsdichte* **F** *puissance volumique* **Pl** *gęstość mocy* **Sv** *effekttäthet*

power distribution *th* • (in a nuclear reactor:) axial and radial variation of the thermal power in the core, a fuel assembly or a fuel element ↦ fuel assembly; fuel element; power-peaking factor **D** *Leistungsverteilung* **F** *distribution de puissance* **Pl** *rozkład mocy* **Sv** *effektfördelning*

power excursion *roc* • very fast increase of the nuclear reactor power above the normal level; the *p.e.* can be either intentional or non-intentional ↦ reactor trip **D** *Reaktorexkursion; Leistungsexkursion* **F** *excursion de puissance* **Pl** *skok mocy; nagły wzrost mocy* **Sv** *reaktorexkursion; effektexkursion*

power-peaking factor *th* • ratio of the local maximum value of the thermal power to its mean value in a reactor core; the *p.-p.f.* can refer to both the heat flux or to the linear power; for reactors with vertical fuel assemblies the *p.-p.f.* can be obtained as a product of the axial, radial and the internal peaking factors ↦ flux-peaking factor; axial peaking factor; radial peaking factor; internal peaking factor **D** *Leistungsformfaktor* **F** *facteur (de forme) de puissance; facteur de point chaud* **Pl** *całkowity współczynnik rozkładu mocy* **Sv** *effektformfaktor; total effektformfaktor*

power range *roc* • power interval of a nuclear reactor within which the power of the reactor, rather than its time constant, is the most important factor for the reactor control ↦ operating range; time-constant range **D** *Leistungsbereich* **F** *domaine de puissance* **Pl** *zakres mocy* **Sv** *effektområde*

power-range monitor *roc* • detection, transmission and recording system for the reactor power; different channels are used for different power ranges; the term is also used, less suitably, to refer to the operating-range monitor ↦ wide range monitor; source range monitor; intermediate range monitor; operating range **D** *Leistungsbereichmonitor* **F** *moniteur du domaine puissance* **Pl** *monitor zakresu mocy* **Sv** *effektkanal; effektmätkanal*

power reactor *rty* • nuclear reactor whose main purpose is energy generating, most frequently in a form of the electric energy ↦ nuclear reactor **D** *Leistungsreaktor* **F** *réacteur de puissance* **Pl**

reaktor mocy; reaktor energetyczny **Sv** *effektreaktor; kraftreaktor*

power stretch *roc* • utilization of a higher than contracted power output in a reactor unit ↦ reactor unit; stretch-out **D** *Erhöhung der Nennleistung* **F** *augmentation de la puissance nominale* **Pl** *zwiększanie mocy nominalnej* **Sv** *effekttöjning*

PRA ↦*probabilistic risk analysis*

Prandtl number *th* • non-dimensional property number for fluid defined as $Pr = \nu/a$, where ν is the kinematic viscosity of the fluid and a is the thermal diffusivity of the fluid ↦ Reynolds number; Nusselt number **D** *Prandtl-Zahl* **F** *nombre de Prandtl* **Pl** *liczba Prandtla* **Sv** *Prandtls tal*

precautionary action zone ↦*emergency planning zone*

precriticality *roc* • the initial phase during the reactor start-up before the criticality is achieved ↦ criticality **D** *Vorkritischer Zustand* **F** *période de construction (avant première criticité)* **Pl** *stan przedkrytycznościowy* **Sv** *förkriticitet*

precursor *rph* • radioactive nuclide which decays into a given, daughter nuclide, either directly or as a later link in a decay chain ↦ nuclide; delayed neutron precursor; decay chain **D** *Vorläufer; Mutternuklid* **F** *précurseur* **Pl** *prekursor; nuklid macierzysty* **Sv** *modernuklid*

preliminary safety analysis report ↦*safety analysis report*

prepressurized fuel *nf* • nuclear fuel in a fuel rod that is filled with high-pressure gas when manufactured to help the cladding to withstand external pressure ↦ fuel rod; cladding; free-standing cladding **D** *Brennstoff mit Vorinnendruck* **F** *combustible sous pression* **Pl** *paliwo pod wstępnym ciśnieniem* **Sv** *trycksatt bränsle*

pressure coefficient of reactivity
nf • reactivity coefficient with respect to pressure in a certain medium or at a certain point in a nuclear reactor ↦ reactivity coefficient **D** *Druckkoeffizient der Reaktivität* **F** *coefficient de pression* **Pl** *ciśnieniowy współczynnik reaktywności* **Sv** *tryckkoefficient*

pressure relief system *nf* • (for boiling water reactor:) system responsible for pressure relief in a reactor vessel through blow-off of steam from the vessel into the containment ↦ boiling water reactor; reactor vessel; containment **D** *Druckentlastung bei Kühlmittelverlust* **F** *circuit de décharge du primaire; système de soufflage[2]* **Pl** *układ dekompresji; układ redukcji ciśnienia* **Sv** *avblåsningssystem*

pressure suppression *rs* • limitation of pressure in a reactor containment at events that lead to pressure increase in the containment such as, e.g., the loss-of-coolant accident; the *p.s.* can be achieved through condensation of the steam or cooling of the gas in water pools, ice beds or pebble beds ↦ containment; loss-of-coolant accident **D** *Druckunterdrückung* **F** *suppression de pression* **Pl** *tłumienie ciśnienia* **Sv** *tryckdämpning; tryckbegränsning*

pressure-suppression chamber ↦*condensation pool*

pressure-suppression pool ↦*condensation pool*

pressure-suppression system *rs* • system for pressure suppression in a reactor containment ↦ pressure suppression; containment **D** *Druckunterdrückungssystem* **F** *système de suppression de pression* **Pl** *układ tłumienia ciśnienia* **Sv** *tryckdämpningssystem; PS-system*

pressure-tube reactor *rty* • reactor in which fuel assemblies are placed inside pressure tubes which sustain the coolant over-pressure ↦ fuel assembly;

coolant **D** *Druckrohrreaktor* **F** *réacteur à tubes de force* **Pl** *reaktor rurowy ciśnieniowy* **Sv** *tryckrörsreaktor*

pressurized heavy-water reactor ↦*PHWR*

pressurized reactor *rty* • nuclear reactor whose primary coolant is under high-enough pressure to prevent boiling, except possibly at hot spot locations; the energy transport in the primary coolant takes place in the liquid phase ↦ primary coolant **D** *Druckreaktor* **F** *réacteur à fluide sous pression* **Pl** *reaktor ciśnieniowy* **Sv** *tryckreaktor*

pressurized-water cooled reactor *rty* • pressurized reactor with water as the primary coolant ↦ pressurized reactor; primary coolant **D** *druckwassergekühlter Reaktor* **F** *réacteur refroidi par eau sous pression* **Pl** *reaktor chłodzony wodą pod ciśnieniem* **Sv** *tryckvattenkyld reaktor*

pressurized water reactor *rty* • pressurized-water cooled reactor with a water moderator ↦ pressurized-water cooled reactor; moderator **D** *Druckwasserreaktor* **F** *réacteur à eau sous pression* **Pl** *reaktor wodny ciśnieniowy* **Sv** *tryckvattenreaktor*

pressurizer *rcs* • component attached with a surge line to the hot leg of the primary coolant system in a pressurized water reactor whose purpose is to keep the pressure in the primary system at the set value ↦ pressurized water reactor **D** *Druckhalter* **F** *pressuriseur* **Pl** *stabilizator ciśnienia* **Sv** *tryckhållningskärl*

pre-stressed concrete pressure vessel *rcs* • reactor pressure vessel made of a pre-stressed concrete ↦ reactor pressure vessel **D** *Druckbehälter aus vorgespanntem Beton* **F** *cuve en béton précontraint* **Pl** *zbiornik ciśnieniowy z betonu sprężonego* **Sv** *betongtrycktank*

primary coolant *th* • coolant in the primary-coolant circuit ↦ primary-coolant circuit; coolant **D** *Primärkühlmittel* **F** *fluide primaire de refroidissement; fluide caloporteur primaire; caloporteur primaire* **Pl** *chłodziwo pierwotne* **Sv** *primärkylmedel*

primary-coolant circuit *th* • coolant circuit with which the thermal power is removed from an energy source such as, e.g., the reactor core ↦ coolant circuit; secondary-coolant circuit; reactor core **D** *Primärkühlkreislauf; Primärkühlkreis* **F** *circuit primaire de refroidissement* **Pl** *pierwotny obieg chłodzenia* **Sv** *primärkylkrets*

primary-coolant system *th* • system containing one or more primary-coolant circuits ↦ primary-coolant circuit **D** *primäres Kühlsystem* **F** *circuit primaire de refroidissement* **Pl** *pierwotny układ chłodzenia* **Sv** *primärkylsystem*

primary system *rcs* • system containing the reactor core, the reactor vessel and the primary-coolant system ↦ primary-coolant system; reactor core; reactor vessel **D** *Primärsystem* **F** *circuit primaire de refroidissement* **Pl** *układ pierwotny* **Sv** *primärsystem*

primary waste *wst* • radioactive wastes in their initial form ↦ secondary waste; radioactive waste **D** *Rohabfall* **F** *déchets primaires* **Pl** *odpad pierwotny* **Sv** *primärt avfall*

principal stresses *mat* • (in solid mechanics:) normal stresses at any point in a solid such as the corresponding tangential stresses are equal to zero ↦ stress corrosion **D** *Hauptspannungen* **F** *contraintes principales* **Pl** *naprężenia główne* **Sv** *huvudspänningar*

PRM ↦*power-range monitor*

probabilistic risk analysis ↦*probabilistic safety analysis*

probabilistic safety analysis *rs* • (abbreviated *PSA:*) probability-based method for analysis of safety of a nuclear installation; for a nuclear reactor installation, a complete *p.s.a.* includes the core damage analysis, the containment analysis, the source term analysis and the analysis of consequences to the environment ↦ nuclear installation; core damage; containment; radioactive source term **D** *probabilistische Sicherheitsanalyse* **F** *analyse probabiliste de la sûreté; évaluation probabiliste des risques* **Pl** *probabilistyczna analiza bezpieczeństwa* **Sv** *probabilistisk säkerhetsanalys; PSA*

probability density function *mth* • function $f_X(x)$ is a *p.d.f.* of the continuous random variable X if for any interval of real numbers $[x_1, x_2]$: (i) $f_X(x) \geq 0$, (ii) $\int_{-\infty}^{\infty} f_X(x)dx = 1$, (iii) $P(x_1 \leq X \leq x_2) = \int_{x_1}^{x_2} f_X(x)dx$ ↦ log-normal distribution; normal distribution; probability distribution **D** *Wahrscheinlichkeightsdichtefunktion* **F** *fonction de densité de probabilité* **Pl** *funkcja gęstości prawdopodobieństwa* **Sv** *sannolikhetstäthetsfunktion*

probability distribution *mth* • *p.d.* of a random variable X is a description of the set of possible values of X, along with the probability associated with each of the possible values ↦ probability density function **D** *Warscheinlichkeitsverteilung* **F** *loi de probabilité* **Pl** *rozkład prawdopodobieństwa* **Sv** *sannolikhetsfördelning*

production reactor *rty* • nuclear reactor which is mainly used for transmutation to produce fissile or other materials, or for irradiation on the industrial scale ↦ fissile material; irradiation **D** *Produktionsreaktor* **F** *réacteur de production* **Pl** *reaktor produkcyjny* **Sv** *produktionsreactor*

promethium *mat* • lanthanide denoted Pm, with atomic number $Z=61$, not naturally abundant; relative atomic mass of longest-lived isotope $A_r=[145]$, density 7.26 g/cm^3, melting point 1042 °C, boiling point 3000 °C; present as a fission product in nuclear reactors, where isotope ^{149}Pm decays into ^{149}Sm, contributing to samarium poisoning of the reactor ↦ samarium poisoning **D** *Promethium* **F** *prométhéum* **Pl** *promet* **Sv** *prometium*

prompt critical *rph* • able to maintain a self-sustained chain reaction without participation of delayed neutrons ↦ delayed neutron; delayed critical **D** *prompt-kritisch* **F** *critique instantané* **Pl** *krytyczny na neutronach natychmiastowych; natychmiastowo-krytyczny* **Sv** *prompt kritisk*

prompt jump ↦ reactor time constant

prompt neutron *rd* • neutron emitted after fission without a measurable delay, typically within about 10^{-14} s; the number ν_p of *p.n.* can vary from 0 to 8 and its average number $\bar{\nu}_p$ is about 2.5, although the precise value depends on the fissile nuclide ↦ fission; delayed neutron **D** *promptes Neutron* **F** *neutron instantané* **Pl** *neutron natychmiastowy* **Sv** *prompt neutron*

prompt neutron fraction *rd* • ratio of the average number of prompt neutrons per fission $\bar{\nu}_p$ and the total (that is prompt and delayed) averaged number of neutrons per fission $\bar{\nu}$ ↦ prompt neutron; fission; delayed neutron fraction **D** *Anteil der prompten Neutronen* **F** *fraction de neutrons instantanés* **Pl** *udział neutronów natychmiastowych* **Sv** *prompt neutronandel; bråkdel prompta neutroner*

protective system *rs* • safety system which shall prevent an operating disturbance from leading to a core accident ↦ core accident **D** *Schutzsystem* **F** *système de protection* **Pl** *układ zabezpieczający* **Sv** *haverimotverkande system*

proton *rd* • stable particle with rest mass $m_p = 1.007\ 276\ 466\ 8$ u and with a positive elementary charge; *p.* is identical with the hydrogen atomic nucleus possessing mass number 1 ↦ atomic mass unit; atomic nucleus; mass number; hydrogen **D** *Proton* **F** *proton* **Pl** *proton* **Sv** *proton*

proton number *bph* • number of protons in the atomic nucleus; for an electrically neutral atom the *p.n.* is equal to the number of electrons ↦ atomic nucleus; atomic number **D** *Protonenzahl* **F** *nombre de protons* **Pl** *liczba protonów* **Sv** *protontal*

prototype reactor *rty* • first reactor in a series of reactors of the same design; sometimes even a smaller reactor, but having the same design features as reactors in the final series is considered as the *p.r.* ↦ reactor **D** *Prototypreaktor* **F** *réacteur prototype* **Pl** *reaktor prototypowy* **Sv** *prototypreaktor*

PSA ↦*probabilistic safety analysis*

PS system ↦*pressure-suppression system*

pulsed column *nf* • (for fuel reprocessing:) column used for solvent extraction in which the water and the organic phases are mixed through liquid oscillations, after which the two phases are separated from each other thanks to the difference in their densities ↦ solvent extraction; fuel reprocessing **D** *Pulskolonne* **F** *colonne pulsée* **Pl** *kolumna pulsacyjna* **Sv** *pulskolonn*

pulsed reactor *rty* • reactor designed to provide intensive neutron bursts for rapid short intervals

through rapid reactivity changes ↦ reactor; reactivity **D** *Pulsreaktor* **F** *réacteur pulsé* **Pl** *reaktor impulsowy* **Sv** *pulsreaktor*

pump-speed control *roc* • control of the speed of main circulation pumps in a boiling water reactor for recirculation control ↦ boiling water reactor; recirculation control **D** *Drehzahlregelung* **F** *régulation de la vitesse des pompes* **Pl** *regulacja prędkości obrotowej pomp* **Sv** *varvtalsreglering*

Purex process *nf* • fuel reprocessing technology in which the solvent extraction is based on tributyl phosphate as the solvent ↦ fuel reprocessing; solvent extraction **D** *Purexprozess* **F** *procédé Purex* **Pl** *technologia Purex* **Sv** *Purex-processen*

PWR ↦*pressurized water reactor*

pyrochemical processing *nf* • (for fuel reprocessing:) procedure based on chemical reactions at a high temperature ↦ fuel reprocessing; pyrometallurgical processing **D** *pyrochemische Behandlung* **F** *traitement pyrochimique* **Pl** *obróbka pirochemiczna* **Sv** *pyrokemisk bearbetning*

pyrometallurgical processing *nf* • (for fuel reprocessing:) procedure based on chemical reactions at high temperature with melt metals, but without a chemical change of the fuel itself ↦ fuel reprocessing; pyrochemical processing **D** *pyrometallurgische Behandlung* **F** *traitement pyrométallurgique* **Pl** *obróbka pirometalurgiczna; obróbka ogniowa* **Sv** *smältmetallurgisk bearbetning*

Qq

Q *gnt* • older unit for a large amount of energy, where Q = 1.06×10^{21} J ↦ energy **D** *Q* **F** *Q* **Pl** *Q* **Sv** *Q*

quality ↦*steam quality*

quality assurance *rs* • all planned and systematic measures necessary to ensure that a product meets given quality requirements ↦ quality control **D** *Qualitätssicherung* **F** *assurance de la qualité* **Pl** *zapewnienie (wysokiej) jakości* **Sv** *kvalitetssäkring*

quality control *rs* • operational methods and activities that are applied to meet the quality requirements ↦ quality assurance **D** *Qualitätsprüfung* **F** *contrôle de la qualité* **Pl** *sterowanie jakością* **Sv** *kvalitetsstyrning*

quality factor *rdp* • (for radiation, denoted *Q*:) factor that indicates how effectiveness of a given absorbed dose depends on the linear energy transfer in water for primary and secondary charged particles; *q.f.* is set according to an agreement ↦ linear energy transfer; dose equivalent; relative biological effectiveness **D** *Bewertungsfaktor* **F** *facteur de qualité* **Pl** *współczynnik jakości (promieniowania)* **Sv** *kvalitetsfaktor*

quality surveillance *rs* • ongoing follow-up and verification of the current state of routines, methods, conditions, processes and products as well as analysis of documented results, in relation to given references, to ensure that specified quality requirements are satisfied in the completed nuclear installation ↦ nuclear installation; quality assurance; quality control **D** *Überwachung der Qualität* **F** *surveillance de la qualité* **Pl** *nadzór nad jakością* **Sv** *kvalitetsövervakning*

Rr

rabbit *rcs* • container for a sample in a pneumatic system ↦ pneumatic system **D** *Rohrpostkapsel* **F** *furet* **Pl** *kapsuła poczty pneumatycznej* **Sv** *rörpostburk*

rad *rdp* • traditional unit for absorbed dose defined as 100 ergs per gram, thus 1 rad = 0.01 gray ↦ absorbed dose; erg; gray^2 **D** *Rad* **F** *rad* **Pl** *rad* **Sv** *rad*

radial peaking factor *th* • (in a reactor with vertical fuel assemblies, also called *radial shape factor*:) ratio of the maximum local power density to the averaged power density over a horizontal cross section in the reactor core ↦ axial peaking factor; power-peaking factor; internal peaking factor **D** *radialer Formfaktor* **F** *facteur (de forme) radial* **Pl** *promieniowy współczynnik rozkładu mocy* **Sv** *radiell formfaktor*

radial shape factor ↦radial peaking factor

radiation *rd* • process of emitting radiant energy as waves or particles; radiation of primary concern in nuclear applications originates in atomic or nuclear processes; it is categorized as follows: charged particulate radiation (fast electrons or heavy charged particles) and uncharged radiation (electromagnetic radiation or neutrons) ↦ ionizing radiation; radiation detector; radiation protection; radiation source **D** *Strahlung* **F** *rayonnement* **Pl** *promieniowanie* **Sv** *strålning*

radiation accident *rdp* • event caused by technical malfunction, incorrect operation or other reasons that results in personal damages due to ionizing radiation ↦ ionizing radiation **D** *Strahlungsunfall* **F** *accident dû aux rayonnements* **Pl** *awaria radiologiczna* **Sv** *strålningsolycka*

radiation chemistry *rdp* • branch of chemistry that deals with chemical effects of ionizing radiation ↦ ionizing radiation; radiochemistry **D** *Strahlenchemie* **F** *chimie sous rayonnement* **Pl** *chemia radiacyjna* **Sv** *strålningskemi*

radiation damage *rdp* • material damage caused by ionizing radiation ↦ ionizing radiation; radiation injury **D** *Strahlenschaden*[1] **F** *dégât par rayonnement* **Pl** *uszkodzenie radiacyjne; uszkodzenie popromienne* **Sv** *strålskada*[1]

radiation detector *mt* • (also called *detector (of radiation)*:) device that indicates or measures radiation where the radiation energy is transformed to an energy form that is more proper for showing or measuring ↦ ionizing radiation; radiation **D** *Strahlungsdetektor* **F** *détecteur de rayonnement* **Pl** *detektor promieniowania* **Sv** *strålningsdetektor; detektor*

radiation heat transfer ↦heat transfer

radiation injury *rdp* • injury of a biologic system caused by ionizing radiation ↦ ionizing radiation; radiation damage **D** *Strahlenschaden*[2] **F** *radiolésion*

Pl *obrażenie popromienne; porażenie radiacyjne* **Sv** *strålskada*[2]

radiation monitor *rdp* ● instrument for measurement of radiation level or activity in order to determine these quantities or to check that they do not exceed the given limiting value ↦ radiation; activity **D** *Strahlungsmonitor; Strahlenwarngerät*[1] **F** *moniteur de rayonnement* **Pl** *monitor promieniowania; przyrząd do kontroli (poziomu) promieniowania* **Sv** *strålningsmonitor; monitor*

radiation physics *rdp* ● branch of physics that deals with properties and physical effects of ionizing radiation ↦ ionizing radiation **D** *Strahlenphysik* **F** *physique des rayonnements* **Pl** *fizyka radiacyjna* **Sv** *strålningsfysik; radiofysik*

radiation protection *rdp* ● (also called *radiological protection* or *health physics*:) activities to limit the effects of ionizing radiation on humans, for example, limiting external exposure to such radiation, limiting the body's absorption of radioactive nuclides and prevention of any harm caused by these two reasons ↦ ionizing radiation; nuclide **D** *Strahlenschutz* **F** *protection contre les rayonnements; radioprotection* **Pl** *ochrona przed promieniowaniem; ochrona radiologiczna* **Sv** *strålskydd*

radiation shielding *rdp* ● branch of technology that deals with shield design ↦ shield **D** *Abschirmungstechnik* **F** *technique de protection (par écrans)* **Pl** *technika osłon przed promieniowaniem* **Sv** *strålskärmsteknik*

radiation sickness *rdp* ● sickness that appears after irradiation of whole or greater part of a body with high doses of ionizing radiation; symptoms include nausea and vomiting, as well as infections and bleeding at a later stage ↦ irradiation; ionizing radiation **D**

Strahlenkrankheit; Strahlenkater **F** *maladie des rayons* **Pl** *choroba popromienna* **Sv** *strålsjuka*

radiation source *rd* ● material or device which emits or can emit ionizing radiation ↦ ionizing radiation **D** *Strahlungsquelle* **F** *source de rayonnement* **Pl** *źródło promieniowania* **Sv** *strålningskälla; strålkälla*

radiation warning assembly *rd* ● instrument that gives an output signal when a certain radiation quantity achieves a previously determined limiting value ↦ radiation monitor **D** *Strahlenwarngerät*[2] **F** *avertisseur de rayonnement* **Pl** *zestaw ostrzegania przed promieniowaniem* **Sv** *strålningsvakt*

radiation work *rdp* ● work that causes occupational exposure ↦ occupational exposure **D** *Arbeit unter Strahlenbelastung* **F** *travail sous rayonnements* **Pl** *praca w warunkach napromieniania* **Sv** *strålningsarbete*

radiative capture *nap* ● capture in an atomic nucleus immediately followed by emission of gamma radiation from the atomic nucleus that was created during the process ↦ capture; atomic nucleus; gamma radiation **D** *Strahlungseinfang* **F** *capture radiative* **Pl** *wychwyt radiacyjny* **Sv** *radiativ infångning*

radioactive *rdy* ● which shows radioactivity or has to do with it ↦ radioactivity; radioactive contamination **D** *radioaktiv* **F** *radioactif* **Pl** *promieniotwórczy; radioaktywny* **Sv** *radioaktiv; aktiv*

radioactive contamination *rdp* ● radioactive substance that is spread in materials or places where it is not desirable ↦ radioactive **D** *radioaktive Kontamination* **F** *contamination radioactive* **Pl** *skażenie promieniotwórcze* **Sv** *radioaktiv kontamination; kontamination*

radioactive decay *rdy* ● nuclear disintegration due to radioactivity ↦ nuclear disintegration; radioactivity **D** *radioaktiver Zerfall* **F** *désintégration radioactive* **Pl** *rozpad promieniotwórczy* **Sv** *radioaktivt sönderfall*

radioactive equilibrium *rdy* ● condition at which activities of the members of a decay chain exponentially decrease with the same radioactive half-lives as that of the precursor; this condition is only possible when the precursor's radioactive half-life is longer than any of the decay products half-lives ↦ decay chain; decay product; precursor; radioactive half-life **D** *radioaktives Gleichgewicht* **F** *équilibre radioactif* **Pl** *równowaga promieniotwórcza* **Sv** *radioaktiv jämvikt*

radioactive fall-out *rdy* ● initially airborne radioactive material that exists in ground air and water or that is deposited on the ground ↦ radioactive material **D** *radioaktiver Niederschlag* **F** *retombées radioactives; retombées* **Pl** *opad promieniotwórczy; opad radioaktywny* **Sv** *radioaktivt nedfall*

radioactive half-life *rdy* ● (usually denoted $T_{1/2}$:) characteristic time during which the number of nuclei of a certain radioactive nuclide decreases by a half through a given type of radioactive decay; the *r.h.-l.* is related to the decay constant λ as follows: $T_{1/2} = \ln 2/\lambda \approx 0.693/\lambda$ ↦ exponential decay; decay constant; effective half-life; biological half-life **D** *radioaktive Halbwertzeit* **F** *période radioactive* **Pl** *okres połowicznego rozpadu; okres połowicznego zaniku; okres półrozpadu; półokres* **Sv** *halveringstid*

radioactive material *rdy* ● material with one or more radioactive components; as far as rules and radiation protection are concerned, the term is limited to materials whose activity or specific activity exceeds a certain value, which includes: (i) mate-

rial designated in national law or by a regulatory body as being subject to regulatory control because of its radioactivity; (ii) any material containing radionuclides where both the activity concentration and the total activity in the consignment exceed the values specified in Transport Regulations; sometimes *radioactive substance* is used instead of *r.m.*; in some terminology *r.m.* ceases to be *r.m.* when it becomes "radioactive waste"; the term "radioactive substance" is used to cover both, i.e., it includes *r.m.* and "radioactive waste" ↦ radioactive waste; activity; specific activity **D** *radioaktiver Stoff* **F** *matière radioactive* **Pl** *materiał promieniotwórczy; substancja promieniotwórcza* **Sv** *radioaktivt material*

radioactive source term *rdp* ● release of a radioactive material from the primary system into the containment (called the internal source term) or from a nuclear installation into the environment (called the external source term) ↦ radioactive material; primary system; containment; nuclear installation **D** *Quellterm in der Radioaktivitätsfreisetzung* **F** *term source* **Pl** *promieniotwórczy człon źródłowy* **Sv** *radioaktiv källterm; källterm*

radioactive substance ↦radioactive *material*

radioactive waste *wst* ● radioactive material obtained through treatment or other handling of nuclear fuel or other radioactive materials and which is considered not to be used for any purpose ↦ radioactive material; nuclear waste **D** *radioaktiver Abfall* **F** *déchets radioactifs* **Pl** *odpad promieniotwórczy* **Sv** *radioaktivt avfall*

radioactivity *rdy* ● property of certain nuclides to spontaneously emit particles (including gamma radiation from the atomic nucleus), or to emit X-rays following electron capture, or to undergo a spontaneous fission ↦ nuclide; gamma radiation; X-ray; fission **D** *Radioaktivität* **F** *radioactivité* **Pl** *promieniotwórczość; radioaktywność* **Sv** *radioaktivitet; aktivitet[2]*

radiobiology *rdp* ● branch of biology that deals with the influence of ionizing radiation on biological systems ↦ ionizing radiation **D** *Strahlenbiologie* **F** *radiobiologie* **Pl** *radiobiologia* **Sv** *strålningsbiologi*

radiochemistry *nch* ● branch of chemistry that deals with radioactive nuclides, their preparation and use ↦ ionizing radiation **D** *Radiochemie* **F** *radiochimie* **Pl** *radiochemia* **Sv** *radiokemi*

radioelement *rdy* ● element that has one or more naturally abundant radioisotopes ↦ element; radioisotope **D** *Radioelement* **F** *radioélément* **Pl** *pierwiastek promieniotwórczy* **Sv** *radioelement*

radiograph[1] *mt* ● apparatus to perform radiography ↦ radiography[2] **D** *Radiographiegerät* **F** *radiographe* **Pl** *radiograf* **Sv** *radiograf*

radiograph[2] *mt* ● picture of an object obtained by interaction of an ionizing radiation with the object or generation of ionizing radiation by the object; the picture can be temporary or permanent and it can be registered in many different ways ↦ radiography[2]; ionizing radiation **D** *Radiogramm* **F** *radiogramme* **Pl** *radiogram* **Sv** *radiogram*

radiography[1] *mt* ● technology to create a radiograph[2] by using ionizing radiation such as X-rays, gamma rays or similar ↦ radiograph[2]; X-ray radiography; gamma radiography **D** *Radiographie[1]* **F** *radiographie[1]* **Pl** *radiografia[1]* **Sv** *radiografi*

radiography[2] *mt* ● creation of a radiograph[2] ↦ radiograph[2] **D**

Radiographie[2] **F** *radiographie[2]* **Pl** *radiografia[2]* **Sv** *radiografering*

radioisotope *rdy* ● radioactive isotope of an element ↦ element; isotope; radionuclide **D** *Radioisotop* **F** *radio-isotope* **Pl** *izotop promieniotwórczy* **Sv** *radioisotop*

radiological physics *rd* ● branch of physics that deals with medical and industrial applications of ionizing radiation; ultraviolet and visible light are usually excluded ↦ ionizing radiation **D** *angewandte Strahlenphysik* **F** *radiophysique; physique radiologique* **Pl** *fizyka radiologiczna* **Sv** *tillämpad strålningsfysik*

radiological protection ↦radiation protection

radiological survey *rdp* ● systematic evaluation of possible radiation risks in connection to certain given conditions as far as usage, release, storage or presence of radiation sources are concerned ↦ radiation source **D** *Strahlenüberwachung* **F** *programme de surveillance radiologique* **Pl** *inspekcja radiologiczna* **Sv** *strålskyddsutredning*

radiology *rd* ● science in which ionizing radiation is used to examine or treat living creatures (medical radiology) or dead objects (technical radiology) ↦ ionizing radiation **D** *Radiologie* **F** *radiologie* **Pl** *radiologia; rentgenologia* **Sv** *radiologi*

radiolysis *nch* ● chemical decomposition caused by ionizing radiation, e.g., high-energy r. of H_2O water molecules into H^+ and OH^- radicals ↦ ionizing radiation **D** *Radiolyse* **F** *radiolyse* **Pl** *radioliza* **Sv** *radiolys*

radiometry *mt* ● measurement technology based on detection of ionizing radiation ↦ ionizing radiation **D** *Radiometrie* **F** *radiométrie* **Pl** *radiometria; pomiary promieniowania* **Sv** *radiometri*

radionuclide *rdy* ● radioactive nuclide (not to be confused with a radioisotope) $\mapsto$ radioisotope **D** *Radionuklid* **F** *radionucléide* **Pl** *nuklid promieniotwórczy; radionuklid* **Sv** *radionuklid*

radium *mat* • radioactive element denoted Ra, with atomic number $Z{=}88$, relative atomic mass of longest-lived isotope $A_r{=}[226]$, density 5.0 g/cm^3, melting point 700 °C, crustal average abundance $9{\times}10^{-7}$ mg/kg and ocean abundance $8.9{\times}10^{-11}$ mg/L; *r.* exists in nature as a decay product of uranium and thorium and can be found in their ores $\mapsto$ element **D** *Radium* **F** *radium* **Pl** *rad* **Sv** *radium*

radon *mat* • radioactive gaseous element denoted Rn, with atomic number $Z{=}86$, relative atomic mass of longest-lived isotope $A_r{=}[222]$, density 4.4 g/cm^3, melting point -71 °C, boiling point -61.7 °C, crustal average abundance $4{\times}10^{-13}$ mg/kg and ocean abundance $6{\times}10^{-16}$ mg/L; *r.* can be found in nearly all types of soil, rock and water; it can migrate into most buildings; studies have linked high concentrations of *r.* to lung cancer $\mapsto$ element **D** *Radon* **F** *radon* **Pl** *radon* **Sv** *radon*

ramp insertion of reactivity *rph* • variation of reactivity in such a way that it changes linearly with time $\mapsto$ reactivity **D** *rampenförmige Reaktivitätserhöhung* **F** *apport linéaire de réactivité* **Pl** *jednostajne zwiększenie reaktywności* **Sv** *reaktivitetsramp*

Rankine cycle *th* • the thermodynamic cycle that is an ideal standard for comparing performance of heat engines, steam power plants and steam turbines that use a condensable vapour as the working fluid $\mapsto$ Brayton cycle; Carnot cycle **D** *Clausius-Rankine-Kreisprozess* **F** *cycle de Rankine* **Pl**

obieg Rankine'a; obieg Clausiusa-Rankine'a **Sv** *Rankine-process*

ratcheting *nf* • successively increasing deformation of the cladding due to repeatedly expanding at power increases and decreases $\mapsto$ cladding **D** *Ratcheting* **F** *rochetage* **Pl** *odkształcenie zapadkowe* **Sv** *hakning*

rated output *roc* • output power included in rating parameters $\mapsto$ rated value; rating **D** *Nennleistung* **F** *puissance nominale* **Pl** *moc znamionowa* **Sv** *märkeffekt*

rated value *roc* • quantity included in rating parameters $\mapsto$ rating **D** *Nennwert* **F** *valeur nominale* **Pl** *wartość znamionowa* **Sv** *märkvärde*

rate meter $\mapsto$ *counting-rate meter*

rate of reactivity *rph* • change of the reactivity per unit time, for example when a control rod is moved or the moderator temperature is changed $\mapsto$ reactivity; control rod; moderator **D** *Reaktivitätsrate* **F** *taux de réactivité* **Pl** *szybkość zmiany reaktywności* **Sv** *reaktivitetsrat*

rating *roc* • values of electrical and mechanical parameters, as well as their duration and time sequence that the manufacturer determined to apply to a machine and which are given on the marking, and which the machine must meet under specified conditions $\mapsto$ rated value **D** *Kennwertung* **F** *régime nominale* **Pl** *dane znamionowe* **Sv** *märkdata*

Rayleigh number *th* • nondimensional number used in natural convection analyses defined as Ra $= \frac{g\beta\Delta T L^3}{a\nu}$ where g is the gravity acceleration, β is the volume expansion coefficient at constant pressure, ΔT is the temperature difference, L is the characteristic length, a is the thermal diffusivity and ν is the kinematic viscosity $\mapsto$ natural convection **D**

Rayleigh-Zahl **F** *nombre de Rayleigh* **Pl** *liczba Rayleigha* **Sv** *Rayleighs tal*

RBMK reactor *rty* • graphite-moderated, boiling water cooled, pressure-tube reactor type, developed in the Soviet Union ↦ pressure-tube reactor **D** *RBMK-Reaktor* **F** *reactéur de type RBMK* **Pl** *reaktor typu RBMK* **Sv** *RBMK-reaktor*

Re ↦*Reynolds number*

reaction rate *nap* • number of nuclear reactions per unit time and usually per unit volume of material; for neutrons with certain energy E, the reaction rate per unit volume $R(\mathbf{r}, E, t)$ at time t and location $\mathbf{r}$ is calculated as a product of neutron flux density $\phi(\mathbf{r}, E, t)$ and a macroscopic cross section at the given energy $\Sigma(E)$, thus $R(\mathbf{r}, E, t) = \Sigma(E)\phi(\mathbf{r}, E, t)$ ↦ neutron flux density; macroscopic cross section **D** *Reaktionsrate* **F** *taux de réaction* **Pl** *szybkość reakcji* **Sv** *reaktionsrat*

reactivity *rph* • (denoted ρ:) measure of deviation from criticality in a system where a chain reaction can take place; r. is calculated as $\rho = (k_{eff} - 1)/k_{eff}$ where k_{eff} is the effective multiplication constant; a positive value of r. indicates a supercritical condition and a negative value a subcritical condition ↦ effective multiplication constant; supercritical; subcritical **D** *Reaktivität* **F** *réactivité* **Pl** *reaktywność* **Sv** *reaktivitet*

reactivity balance *rph* • list of positive and negative contributions to the reactivity of a nuclear reactor ↦ reactivity **D** *Reaktivitätsbilanz* **F** *bilan de réactivité* **Pl** *bilans reaktywności* **Sv** *reaktivitetsbudget*

reactivity coefficient *rph* • partial derivative of reactivity with respect to a characteristic quantity in a nuclear reactor ↦ power coefficient of reactivity; pressure coefficient of reactivity; temperature coefficient of reactivity; void coefficient of reactivity **D** *Reaktivitätskoeffizient* **F** *coefficient de réactivité* **Pl** *współczynnik reaktywności* **Sv** *reaktivitetskoefficient*

reactivity feed-back *rph* • influence of changes in certain operating parameters of a nuclear reactor on reactivity; operating parameters that are usually considered include the reactor power, the fuel and moderator temperatures, the system pressure and the void fraction in the core ↦ reactivity; void fraction **D** *Reaktivitätrückkoplung* **F** *effet rétroactif sur la réactivité* **Pl** *reaktywnościowe sprzężenie zwrotne* **Sv** *reaktivitetsåterkoppling*

reactivity-induced accident *rs* • (abbreviated *RIA:*) core accident that is caused by an uncontrolled reactivity increase; the *r.-i. a.* can be caused, for example, by a wrong control-rod manoeuvre during reactor power increase or during reactor start-up ↦ core accident; reactor start-up **D** *Reaktivitätsunfall* **F** *accident de réactivité* **Pl** *awaria reaktywnościowa* **Sv** *reaktivitetshaveri*

reactivity meter *roc* • electronic system that solves the reactor kinetics equations with respect to reactivity changes, using the corresponding power changes as the input signal ↦ reactivity **D** *Reaktivitätsmesser* **F** *réactimètre* **Pl** *miernik reaktywności* **Sv** *reaktivitetsmätare*

reactor *rty* • (also called *nuclear reactor:*) installation in which a controllable chain reaction of nuclear processes can be maintained; usually the nuclear fission process is considered ↦ nuclear chain-reaction; nuclear fission **D** *Reaktor* **F** *réacteur* **Pl** *reaktor; reaktor jądrowy* **Sv** *reaktor; kärnreaktor*

reactor cavity *rcs* • (for pressurized water reactor:) space under and around a reactor vessel, inside the biological shield ↦ pressurized water reactor; reactor vessel; biological shield **D** *Reaktorgrube* **F** *piscine du réacteur* **Pl** *wnęka reaktor* **Sv** *reaktorgrop*

reactor cell[1] *rph* ● one of many elementary regions in a heterogeneous reactor, which all have the same geometrical shape and neutron-physical properties ↦ heterogeneous reactor; neutron physics **D** *Reaktorzelle*[1] **F** *cellule de réacteur*[1] **Pl** *komórka reaktora*[1] **Sv** *cell*

reactor cell[2] *rph* ● cell around the fuel assembly in the reactor core ↦ fuel assembly **D** *Reaktorzelle*[2] **F** *cellule de réacteur*[2] **Pl** *komórka reaktora*[2] **Sv** *reaktorcell; gittercell*

reactor control[1] *roc* ● automatic change of the reaction rate in a reactor or an adjustment of its reactivity to maintain or achieve the desired operational mode ↦ reactor control[2]; reaction rate; reactivity **D** *Reaktorregelung* **F** *commande d'un réacteur*[1] **Pl** *regulacja reaktora* **Sv** *reaktorreglering*

reactor control[2] *roc* ● intentional change of the reaction rate in a reactor or an adjustment of its reactivity to maintain or achieve the desired operational mode ↦ reactor control[1]; reaction rate; reactivity **D** *Reaktorsteuerung* **F** *commande d'un réacteur*[2] **Pl** *sterowanie reaktorem* **Sv** *reaktorstyrning*

reactor control system *roc* ● system comprising the components and materials that cooperate during reactor control ↦ reactor control[1]; reactor control[2] **D** *Reaktorsteuersystem* **F** *système de conduite d'un réacteur* **Pl** *układ sterowania reaktorem* **Sv** *reaktorstyrsystem; styrsystem*

reactor core *rcs* ● 1. part of a nuclear reactor where the nuclear fission takes place; in a thermal reactor, the *r.c.*

contains fuel, moderator, and coolant; in a fast reactor core, only fuel and coolant are present, whereas a fast breeder reactor core can also contain a fertile material 2. (in breeder reactor:) part of the nuclear reactor initially containing the fissile material ↦ reactor; fission; fuel; moderator; coolant; breeder reactor; fissile material; fertile material **D** *Spaltzone; Reaktorkern* **F** *coeur d'un réacteur; coeur* **Pl** *rdzeń reaktora* **Sv** *reaktorhärd; härd*

reactor doubling time ↦*doubling time*[1,2,3]

reactor engineering *gnt* ● part of nuclear technology that deals with the nuclear reactor construction ↦ nuclear technology; reactor **D** *Reaktortechnik* **F** *technique des réacteurs nucléaires* **Pl** *inżynieria reaktorowa* **Sv** *reaktorteknik*

reactor excursion ↦*power excursion*

reactor lattice *nf* ● array of fuel assemblies and other components arranged in a regular pattern in the reactor core; two dominant types of *r.l.* are the square lattice (employed in PWRs and BWRs) and the hexagonal lattice (employed in HTGRs and sodium-cooled fast reactors) ↦ fuel assembly; reactor core **D** *Reaktorgitter* **F** *réseau de réacteur* **Pl** *siatka reaktora* **Sv** *reaktorgitter*

reactor loop *rcs* ● pipeline in a nuclear reactor, through which a liquid or gas can pass as part of the reactor operation or for experimental purposes ↦ reactor **D** *Reaktorversuchskreislauf* **F** *boucle de réacteur* **Pl** *pętla reaktora* **Sv** *reaktorslinga*

reactor noise *rph* ● fluctuations of the neutron flux density and thus of the reactor power that occur in a nuclear reactor, caused by the random fluctuations of nuclear processes in the core, or random variations in the mechanical or hydrodynamic processes that affect the reactivity ↦ neutron flux density; reactor core; reactivity **D** *Reaktorrauschen* **F** *bruit d'un réacteur* **Pl** *szum reaktora* **Sv** *reaktorbrus*

reactor operator *roc* • engineer who is a member of the control-room personnel ↦ control room; control-room personnel **D** *Reaktoroperateur* **F** *exploitant d'un réacteur* **Pl** *operator reaktora* **Sv** *kontrollrumsingenjör*

reactor oscillator *mt* • device which provides periodic variations of the neutron flux density in a nuclear reactor designed to measure, for example, neutron-absorption properties of the nuclear reactor, or of a certain body ↦ reactor; neutron flux density **D** *Reaktoroszillator* **F** *oscillateur de pile* **Pl** *oscylator reaktorowy* **Sv** *reaktoroscillator; reaktormodulator*

reactor period ↦*reactor time constant*

reactor physics *rph* • branch of physics that deals with the physical principles of the mode of operation of nuclear reactors, especially neutron distribution in space, time and energy ↦ reactor; neutron flux density **D** *Reaktorphysik* **F** *physique des réacteurs* **Pl** *fizyka reaktorów* **Sv** *reaktorfysik*

reactor pressure vessel *rcs* • (abbreviated *RPV*:) reactor vessel adapted to withstand considerable working pressure; the working pressure is different for different types of reactors and is around 7 MPa for BWRs and 17 MPa for PWRs ↦ reactor vessel; reactor **D** *Reaktordruckbehälter; Druckbehälter; Reaktordruckgefäß; Druckgefäß* **F** *caisson résistant d'un réacteur* **Pl** *zbiornik ciśnieniowy reaktora; zbiornik ciśnieniowy* **Sv** *reaktortrycktank; trycktank*

reactor simulator *roc* • computer-based equipment that is used to simulate transient processes in nuclear reactors ↦ nuclear power plant simulator **D** *Reaktorsimulator* **F** *simulateur de réacteur* **Pl** *symulator reaktora* **Sv** *reaktorsimulator*

reactor stability *roc* • reactor operation property reflecting its power response to any direct or indirect reactivity perturbation; the *r.s.* can be analysed using various tools, such as linear stability analysis, non-linear time-domain stability analysis or Lyapunov stability theory ↦ reactivity; Lyapunov stability **D** *Stabilität des Reaktors* **F** *stabilité du réacteur* **Pl** *stabilność reaktora* **Sv** *reaktorstabilitet*

reactor start-up *roc* • increase of reactivity in the reactor until it becomes critical and its power increases in a controlled manner ↦ reactivity; critical **D** *Anfahren des Reaktors* **F** *démarrage d'un réacteur* **Pl** *włączenie reaktora; uruchomienie reaktora* **Sv** *reaktorstart*

reactor technology *gnt* • science that deals with nuclear engineering ↦ nuclear engineering **D** *Reaktortechnologie* **F** *technologie de réacteurs nucléaires* **Pl** *technologia reaktorów jądrowych* **Sv** *reaktortechnologi*

reactor time constant *roc* • (also called *reactor period*:) time T required to change the neutron flux density by a factor $e = 2.718...$ when the flux exponentially increases or decreases with time as $\phi(t) = \phi_0 e^{t/T}$, where ϕ_0 is the neutron flux density at time $t = 0$; the *stable reactor period* is the reactor period when the very rapid initial transient caused by prompt neutrons (called the *prompt jump*) is neglected; using a point reactor kinetics model with one effective delayed group, the stable reactor period is $T \cong (\beta - \rho_0)/\lambda\rho_0$, where β - delayed neutron fraction, λ - averaged decay constant of the precursors of the delayed neutrons, ρ_0 - reactivity change

↦ neutron flux density; delayed neutron fraction; reactivity **D** *Reaktorzeitkonstante; Reaktorperiode* **F** *constante de temps d'un réacteur* **Pl** *stała czasowa reaktora; okres asymptotyczny reaktora* **Sv** *reaktortidkonstant*

reactor trip *roc* ● (replaces the term *scram*, which dates from early days of reactor development and is an acronym for *S*afety *C*ontrol *R*od *A*xe *M*an:) rapid reduction of nuclear reactor power by making it sub-critical; the *r.t.* can be triggered automatically or manually; automatic triggering can occur intentionally or unintentionally when the reactor protection system comes into operation ↦ subcritical **D** *Schnellabschaltung* **F** *arrêt d'urgence* **Pl** *awaryjne wyłączenie reaktora* **Sv** *snabbstopp*

reactor unit *gnt* ● plant for production of electricity consisting of a nuclear reactor, a turbine and a generator ↦ reactor **D** *Reaktorblock* **F** *tranche (d'une centrale)* **Pl** *blok reaktora* **Sv** *reaktorblock; kärnkraftblock*

reactor vessel *rcs* ● vessel that encloses a nuclear reactor core and usually also its reflector; in a boiling water reactor, the *r.t.* also includes, among others, steam separators and steam dryers ↦ reactor; reflector; boiling water reactor **D** *Reaktorbehälter* **F** *caisson du réacteur* **Pl** *zbiornik reaktora* **Sv** *reaktortank*

reactor with external circulation *rty* ● boiling water reactor where the coolant is circulated by pumps arranged outside the reactor pressure vessel ↦ boiling water reactor; coolant; reactor pressure vessel **D** *Reaktor mit externen Kühlmittelpumpen* **F** *réacteur avec circulation externe* **Pl** *reaktor z zewnętrznym obiegiem* **Sv** *externpumpsreaktor*

reactor with internal main recir-

culation *rty* ● boiling water reactor where coolant is circulated by pumps located inside the reactor pressure vessel ↦ boiling water reactor; coolant; reactor pressure vessel **D** *Reaktor mit internen Kühlmittelpumpen* **F** *réacteur avec pompes de circulation internes* **Pl** *reaktor z wewnętrznymi pompami recyrkulacyjnymi* **Sv** *internpumpsreaktor*

recirculation *roc; rs* ● **1.** normal mode of operation for cooling of a boiling water reactor core by circulation of the water content in the core **2.** special operating mode during the emergency core cooling, following the loss-of-coolant accident in a pressurized water reactor, wherein the exiting water is cooled and returned to the nuclear reactor primary cooling system ↦ boiling water reactor; pressurized water reactor; emergency core cooling; loss-of-coolant accident; reactor core **D** *Rezirkulation* **F** *recirculation* **Pl** *recyrkulacja* **Sv** *recirkulation*

recirculation control *roc* ● (during reactor operation:) controlling the power of a boiling water reactor by changing the coolant flow which, via the void coefficient, affects the reactivity ↦ reflector control; boiling water reactor; coolant; void coefficient of reactivity; reactivity **D** *Pumpensteuerung* **F** *contrôle par vide* **Pl** *regulacja recyrkulacją chłodziwa* **Sv** *kylflödesstyrning; pumpstyrning*

recirculation internal pump ↦ *internal primary recirculation pump*

redundancy *rs* ● excess in order to increase the reliability of components or systems in a structure, e.g., in a cooling system ↦ emergency core cooling system **D** *Redundanz* **F** *redondance* **Pl** *redundancja* **Sv** *redundans*

reference level for emergency action *rdp* ● (in emergency preparedness:) reference for grading action at various abnormal events; an *emergency action level (EAL)* is defined as a specific, predetermined, observable criterion (such as, e.g., an instrument reading) used to detect, recognize and determine the *emergency class*; three emergency classes (in order of increasing severity) are defined: *alert* - following an event of unknown or significant decrease in the level of protection for the public or on-site personnel; *site area emergency* - following an event resulting in a major decrease in the level of protection for the public or on-site personnel; *general area emergency* - following an event resulting in an actual release, or substantial probability of a release, requiring implementation of urgent protective actions off-site ↦ emergency preparedness **D** *Referenzniveau für Notmaßnahmen* **F** *niveau de référence en cas d'urgence* **Pl** *poziom odniesienia dla postępowania awaryjnego* **Sv** *larmnivå*

refill *th* • stage of the loss-of-coolant accident during which the emergency coolant water begins filling the lower plenum of reactor vessel up to the bottom of the core ↦ loss-of-coolant accident; reflooding **D** - **F** - **Pl** *ponowne napełnianie zbiornika reaktora* **Sv** *återfyllning*

reflector *rcs* • material or body which reflects incident radiation; in nuclear reactor technology, the *r.* commonly refers to a reactor part located adjacent to the reactor core, with the task to return the neutrons to the core ↦ reactor; reactor core; neutron **D** *Reflektor* **F** *réflecteur* **Pl** *reflektor* **Sv** *reflektor*

reflector control *roc* • reactor control by changing the properties of the reflector, its position or size ↦ reactor control[2]; reflector **D** *Reflektorsteuerung* **F** *commande par le réflecteur* **Pl** *sterowanie reflektorem* **Sv** *reflektorstyrning*

reflector saving *rph* • saving in reactor core dimensions, critical mass, etc., that with unchanged reactivity, can be done by a reflector placed around the reactor core ↦ reactor core; critical mass; reflector **D** *Reflektorersparnis* **F** *économie due au réflecteur* **Pl** *oszczędność reflektorowa* **Sv** *reflektorbesparing*

reflooding *th* • stage of loss-of-coolant accident during which the water level would rise sufficiently to cool the core ↦ reactor vessel; emergency core cooling; loss-of-coolant accident; refill **D** *Einspeisung der Notkühlung* **F** *renoyage* **Pl** *ponowne zalewanie rdzenia* **Sv** *flödning*

refueling *nf* • removal of spent fuel and an introduction of new nuclear fuel in a nuclear reactor ↦ spent fuel **D** *Brennstofferneuerung* **F** *renouvellement du coeur; rechargement* **Pl** *uzupełnianie paliwa* **Sv** *bränslebyte*

RELAP *th* • (acronym for Reactor Excursion and Leak Analysis Program:) tool for analysing accidents and system transients in nuclear power plants; it can model thermal-hydraulic phenomena in connected volumes that represent vessels and pipelines ↦ TRACE **D** *RELAP* **F** *RELAP* **Pl** *RELAP* **Sv** *RELAP*

relative atomic mass *bph* • (denoted A_r:) ratio of the atom's mass to that of one-twelfth of a neutral atom of ^{12}C in its ground state ↦ atomic mass; atomic mass unit **D** *relative Atommasse; Atomgewicht* **F** *masse atomique relative; poids atomique* **Pl** *względna masa atomowa* **Sv** *relativ atommassa*

relative biological effectiveness *rdp* • (for a given living organism or a portion thereof:) ratio of the absorbed dose of a reference radiation which provides a specific biological effect, and the absorbed dose of the current radiation that has the same ef-

fect (only in radiation biology) ↦ absorbed dose; dose equivalent; radiation **D** *relative biologische Wirksamkeit* **F** *efficacité biologique relative* **Pl** *względna efektywność biologiczna* **Sv** *relativ biologisk effektivitet*
relative conversion ratio *rph* • ratio of the instantaneous conversion ratio in a nuclear reactor and the instantaneous conversion ratio of fuel of the same composition but in a specified (usually thermal) neutron energy spectrum; the *r.c.r.* is easier to determine experimentally than the absolute conversion ratio ↦ conversion ratio **D** *relatives Konversionsverhältnis* **F** *rapport de conversion relatif* **Pl** *względny współczynnik konwersji* **Sv** *relativt konversionsförhållande*

relative importance *rph* • (neutron type A relative to neutrons type B:) average number of neutrons of the velocity and position B, which must be supplied to a critical system for it to remain exactly critical, after one neutron of the velocity and position A is removed ↦ critical; neutron importance function **D** *relativer Einfluß* **F** *importance relative* **Pl** *względna cenność* **Sv** *relativ importans*

rem *rdp* • (acronym for *r*öntgen *e*quivalent *m*an:) older unit of the dose equivalent and the effective dose equivalent where 1 rem = 0.01 Sv; ↦ dose equivalent; effective dose equivalent; sievert; roentgen **D** *Rem* **F** *rem* **Pl** *rem* **Sv** *rem*

remote maintenance *roc* • supervision and maintenance of a radioactive or radioactive contaminated equipment with the use of remote controlled devices ↦ radioactive; radioactive contamination **D** *Fernbedienung* **F** *maintenance à distance* **Pl** *zdalna konserwacja* **Sv** *fjärrtillsyn*

removal cross section *xr* • (in ra-

diation protection:) effective neutron cross section which is assigned to a material placed between a source of fission neutrons and a thick layer of hydrogen-containing material; the *r.c.s.* is used for calculation of the relaxation length of the flux density of fast neutrons in thick screens ↦ radiation protection; cross section; neutron source; fission neutron **D** *Ausscheidequerschnitt; Removalquerschnitt* **F** *section efficace de déplacement* **Pl** *przekrój czynny na usuwanie* **Sv** *svinntvärsnitt*[2]

removal-diffusion theory *rdp* • theory for calculating neutron attenuation in some shielding materials, where the attenuation process is treated in two stages: in the first stage a source of neutrons that collided once is determined using the removal cross sections; in the second stage, the neutrons found in stage one are treated with diffusion theory ↦ attenuation; removal cross section; diffusion theory; shield **D** *Removal-Diffusionstheorie* **F** *théorie combinant déplacement et diffusion* **Pl** *teoria dyfuzji z usuwaniem* **Sv** *svinndiffusionsteori*

replacement of the whole charge *nf* • replacement of all nuclear fuel in the reactor core at one time ↦ reactor core **D** *vollständiger Austausch der Ladung* **F** *remplacement complet de la charge* **Pl** *wymiana całego rdzenia* **Sv** *helhärdsbyte*

repository ↦*ultimate storage*

research reactor *rty* • nuclear reactor used mainly as a tool for basic and applied research ↦ irradiation reactor; materials testing reactor; training reactor **D** *Forschungsreaktor* **F** *réacteur de recherche* **Pl** *reaktor badawczy* **Sv** *forskningsreaktor*

residual heat removal *rs* • core cooling in a shutdown nuclear reactor ↦ residual power; decay power **D** *Nachwärmekühlung* **F** *refroidissement* du réacteur à l'arrêt **Pl** odbiór ciepła powyłączeniowego **Sv** resteffektkylning

residual power *th* • power output of a nuclear reactor after shutdown resulting from the residual radioactivity and the accumulated heat ↦ residual heat removal; decay power **D** Nachleistung **F** puissance résiduelle **Pl** moc powyłączeniowa **Sv** eftereffekt

resonance absorption of neutrons *xr* • neutron absorption in the resonance energy range ↦ resonance energy; resonance escape probability; resonance capture of neutrons **D** Neutronenresonanzabsorption; Resonanzabsorption **F** absorption de neutrons par résonance **Pl** absorpcja rezonansowa neutronów **Sv** resonansabsorption

resonance capture of neutrons *xr* • radiative capture of neutrons in the resonance energy range ↦ resonance energy; resonance escape probability; resonance absorption of neutrons **D** Neutronenresonanzeinfang; Resonanzeinfang **F** capture de neutrons de résonance **Pl** wychwyt rezonansowy neutronów **Sv** resonansinfångning

resonance detector *mt* • activation detector whose cross section for neutrons is characterized by one or more large resonances and therefore it provides information about the neutron flux density in the resonance energy range ↦ resonance energy; detector **D** Resonanzdetektor **F** détecteur résonnant **Pl** detektor rezonansowy **Sv** resonansdetektor

resonance energy *xr* • kinetic energy (measured in the laboratory system) of a particle incident to an atomic nucleus such that the total energy of the system which consists of the particle and the nucleus is nearly equivalent to the energy level of a *compound nucleus*, i.e., a strongly excited nucleus with short life, which is formed as an intermediate stage in the nuclear reaction ↦ resonance absorption

of neutrons; resonance capture of neutrons **D** Resonanzenergie **F** énergie de résonance **Pl** energia rezonansu **Sv** resonansenergi

resonance escape probability *rph* • probability that a neutron with a certain energy, through slowing down without leakage, will achieve a lower energy without being absorbed in the resonance range ↦ resonance energy; resonance absorption of neutrons **D** Resonanzentkommwahrscheinlichkeit **F** facteur antitrappe **Pl** prawdopodobieństwo uniknięcia wychwytu rezonansowego **Sv** resonanspassagefaktor

resonance integral *rph* • integral with respect to the energy of the product of a substance absorption cross section and a reciprocal neutron energy, calculated over the resonant region or its fraction; if the neutron flux density is inversely proportional to the energy, the resonance integral gives a measure of the probability of neutron absorption between the integration limits; if the neutron flux density does not vary in this way, the effective resonance integral is used; the term may sometimes be specified to apply to a certain reaction such as capture, fission or activation ↦ resonance energy; resonance escape probability; resonance absorption of neutrons **D** Resonanzintegral **F** intégrale de résonance **Pl** całka rezonansowa **Sv** resonansintegral

resonance neutron *rd* • neutron with energy in the resonance region ↦ resonance energy; resonance region; neutron **D** Resonanzneutron **F** neutron de résonance **Pl** neutron rezonansowy **Sv** resonansneutron

resonance region *xr* • neutron energy range within which cross sections for many nuclides exhibit peaks caused by resonance energy levels ↦ resonance energy; neutron **D** Resonanzgebiet **F** *région de résonance; zone de réso-nance* **Pl** *obszar rezonansowy; strefa rezo-nansowa* **Sv** *resonansområde*

response matrix method *rph* • general method for solution of parti-cle transport problems, wherein each subspace is described by a response matrix; response matrix method is used, between others, to formulate equations in the nodal method ↦ neutron transport equation; nodal method **D** *Responsmatrix-Methode* **F** *méthode de la matrice de réponse* **Pl** *metoda macierzy odpowiedzi* **Sv** *responsmatrismetoden*

retrofitting *roc* • change of build-ings, systems or components of a nu-clear installation; additional require-ments set by the authority can be a reason for the adjustment ↦ nuclear installation **D** *Nachrüstung; nachträgliche Ausrüstung* **F** *rattrapage; modification en cours d'exploitation* **Pl** *modyfikowanie pod-czas eksploatacji* **Sv** *efterjustering*

reversible process *th* • idealized thermodynamic process without fric-tion or other losses whose direction can be reversed ↦ Carnot cycle; irre-versible process **D** *reversibler Prozess* **F** *pro-cessus réversible* **Pl** *proces odwracalny* **Sv** *reversibel process*

revision period ↦*overhaul period*

rework cell *nf* • device for adjust-ing the properties of radioactive prod-ucts from a chemical process that does not fulfil given specifications; proper-ties that need to be adjusted can be, for example, the concentration level or the pH value ↦ fuel reprocessing **D** *Ein-stellzelle* **F** *atelier de recyclage* **Pl** *komora przerobowa* **Sv** *återbearbetningscell*

Reynolds number *th* • (in fluid me-chanics, denoted Re:) dimensionless quantity that is equivalent to the ra-tio of inertial forces to viscous forces; for flow of fluid with density ρ, veloc-ity U, dynamic viscosity μ and with characteristic linear dimension L, the *R.n.* is defined as, Re= $\rho U L/\mu$; in all viscous flows *R.n.* is the primary con-trolling parameter; in particular, for a given geometry, as the *R.n.* increases, the flow pattern changes from laminar through a transitional region into the turbulent region ↦ laminar flow; turbu-lent flow **D** *Reynolds-Zahl* **F** *nombre de Reynolds* **Pl** *liczba Reynoldsa* **Sv** *Reynolds tal*

RIA ↦*reactivity-induced accident*

RIP ↦*recirculation internal pump*

rock depository ↦*bedrock depository*

roentgen *rdp* • (denoted R:) old unit of exposure in which 1 R = 2.58 × 10^{-4} C/kg (exactly); the *r.* is cur-rently superseded by the SI unit C/kg ↦ exposure **D** *Röntgen* **F** *röntgen* **Pl** *rönt-gen* **Sv** *röntgen*

Rossi alpha *rph* • inverse of the time constant that would be obtained for a nuclear reactor if no delayed neutrons were emitted ↦ reactor time constant; de-layed neutron **D** *Abklingkonstante* α **F** *al-pha de Rossi* **Pl** *alfa Rossiego* **Sv** *Rossi-alfa*

RPV ↦*reactor pressure vessel*

running down *roc* • (of a turbine set or a pump:) reducing the rotational speed from the rated speed to the standstill; if the turning equipment is started before a complete stop, the ongoing *r.d.* continues until the turn-ing starts ↦ running up; turning **D** *Ab-fahren* **F** *réduction graduelle de la puissance; réduction de la puissance; baisse progressive de puissance* **Pl** *wybieg* **Sv** *utrullning*

running-in period *roc* • period that follows the first charge of a nuclear reactor with a fresh fuel and dur-ing which appropriate shape factors are obtained by choosing the fuel and absorber distribution that may be changed when the nuclear fuel reaches a certain degree of the burnup and it is gradually replaced ↦ running-out period; equilibrium cycle *D Einlaufzeit F période de début de vie Pl okres docierania Sv inkörningsperiod*

running-out period *roc* • the period at the end of a reactor lifetime during which the last charge of fuel is consumed and the nuclear reactor is prepared for the final shutdown ↦ running-in period *D Auslaufzeit F période de fin de vie Pl okres końcowy eksploatacji Sv slutkörningsperiod*

running up *roc* • (of a turbine set or a pump:) increasing the rotation speed from the standstill to the rated speed; if the turning of the equipment is ongoing, the term *r.u.* means increasing the rotation speed from the turning speed to the rated speed ↦ running down; turning *D Anfahren F augmentation de puissance Pl rozbieg Sv upprullning*

Ss

safeguards (of nuclear materials) *sfg* ● checking that the nuclear material is used for peaceful purposes only ↦ nuclear material **D** *Überwachung des Kernmaterials* **F** *contrôle des matières nucléaires* **Pl** *zabezpieczenie (materiałów jądrowych)* **Sv** *kärnämneskontroll*

safe mass *rph* ● (for a given fissile material:) minimum critical mass divided by a safety factor greater than unity ↦ fissile material; critical mass **D** *sichere Masse* **F** *masse sûre* **Pl** *masa bezpieczna* **Sv** *kriticitetssäker massa*

safe shutdown earthquake *rs* ● (abbreviated *SSE*:) largest earthquake a nuclear installation must be able to endure in such a way that it can be shut down and maintained in a safe condition ↦ earthquake; nuclear installation **D** *Auslegungserdbeben[2]* **F** *séisme majoré de sécurité* **Pl** *graniczne trzęsienie ziemi dla bezpiecznego wyłączenia reaktora* **Sv** *gränssättande jordskalv för säker avställning*

safety analysis report *rs* ● (abbreviated *SAR*:) document that must be submitted by an applicant to the licensing authority for a permit to construct and later for a license to operate a nuclear power plant; one section of the *s.a.r.* consists of analyses of a wide range of conceivable abnormal events with the purpose to show that the plant is designed in such a manner that these events can be con-

trolled or that their consequences are accommodated without undue risk to the health and safety to the public; a *preliminary safety analysis report* (PSAR) is submitted to obtain the construction permit, and a *final safety analysis report* (FSAR) is submitted to obtain an approval of the final plant design ↦ nuclear power plant **D - F - Pl** *raport analizy bezpieczeństwa* **Sv** *säkerhetsrapport*

safety chain *rs* ● group of safety circuits for triggering a certain safety function; each *s.ch.* is divided into redundant channels; the signal from at least two channels is required for triggering a safety function ↦ safety circuit; safety function **D** *Sicherheitskreis* **F** *chaîne de circuits de sécurité* **Pl** *łańcuch bezpieczeństwa* **Sv** *säkerhetskedja*

safety circuit *rs* ● (for a nuclear reactor:) logic circuit which receives information from different measurements in a nuclear reactor and which, on the basis thereof, automatically triggers one or more security functions ↦ safety chain; safety function **D** *Reaktorschutzsystem* **F** *circuit de sécurité* **Pl** *obwód bezpieczeństwa* **Sv** *säkerhetskrets*

safety element *rs* ● (also called *safety member*:) control element that provides a reserve of negative reactivity for the reactor trip; the *s.e.* can work alone or in groups ↦ safety chain; safety circuit; safety function; reactor trip

D *Sicherheitselement* **F** *élément de sécurité* **Pl** *element bezpieczeństwa* **Sv** *säkerhetselement*

safety function *rs* • action of one or more safety systems; the *s.f.* can be, e.g., the reactor trip, the isolation or the emergency core cooling ↦ safety system; reactor trip; isolation; emergency core cooling **D** *Sicherheitsfunktion* **F** *fonction de sûreté* **Pl** *funkcja bezpieczeństwa* **Sv** *säkerhetsfunktion*

safety injection *rs* • supply of the primary coolant from the emergency cooling of the reactor core when the normal cooling is insufficient, e.g., due to breakage or leakage in the primary system; the term is mainly used in PWRs ↦ primary coolant; primary system **D** *Sicherheitseinspeisung* **F** *injection de sécurité* **Pl** *wtrysk bezpieczeństwa* **Sv** *säkerhetsinsprutning*

safety member ↦*safety element*

safety rod *rs* • rod-shaped safety element ↦ safety element; cluster control rod **D** *Sicherheitsstab* **F** *barre de sécurité* **Pl** *pręt bezpieczeństwa; pręt awaryjny* **Sv** *säkerhetsstav*

safety system *rs* • system designed to maintain the safety of a nuclear installation in case of abnormal events ↦ nuclear installation; protective system; mitigative system **D** *Sicherheitssystem* **F** *système de sûreté* **Pl** *układ bezpieczeństwa* **Sv** *säkerhetssystem*

samarium *mat* • chemical element denoted Sm, with atomic number $Z=62$, relative atomic mass $A_r=150.36$, density 7.52 g/cm^3, melting point 1074 °C, boiling point 1794 °C, crustal average abundance 7.05 mg/kg and ocean abundance 4.5×10^{-7} mg/L; isotope ^{149}Sm is a nuclear poison since it is a fission product with large fission yields and a significant absorption cross section

↦ element; nuclear poison; samarium poisoning **D** *Samarium* **F** *samarium* **Pl** *samar* **Sv** *samarium*

samarium poisoning *rph* • forming of a fission product ^{149}Sm in a thermal reactor; the isotope ^{149}Sm is a nuclear poison ↦ samarium; thermal reactor; fission product; nuclear poison **D** *Samariumvergiftung* **F** *empoisonnement samarium* **Pl** *zatrucie samarem* **Sv** *samariumförgiftning*

SAR ↦*safety analysis report*

saturated boiling *th* • boiling in a channel where the cross-section mean enthalpy of the two-phase mixture is equal to or greater than the saturation enthalpy of the liquid phase at the local pressure ↦ nucleate boiling; subcooled boiling; film boiling **D** *Blasensieden*^{2} **F** *ébullition saturée; ébullition en masse; pleine ébullition* **Pl** *wrzenie nasycone* **Sv** *mättnadskokning; volymkokning*

saturation activity *rdy* • maximum activity that can be obtained from a nuclide by activation of a sample of the material in a given flux density ↦ activity; nuclide **D** *Sättigungsaktivität* **F** *activité à saturation* **Pl** *aktywność nasycenia* **Sv** *mättnadsaktivitet*

SBO ↦*station blackout*

scaler *mt* • device for counting electric pulses, containing one or more counter circuits ↦ counting rate; counting-rate meter; counter; count **D** *Zähler*^{2} **F** *échelle de comptage* **Pl** *przelicznik* **Sv** *pulsräknare*

scattering *rph* • process wherein a direction of motion or energy of a particle change because of a collision with another particle or with a particle system ↦ scattering cross section; scattering law; scattering kernel **D** *Streuung* **F** *diffusion* **Pl** *rozpraszanie* **Sv** *spridning*^{2}

scattering cross section *xr* • cross section for the scattering process ↦

scattering; scattering kernel; scattering law **D** *Streuquerschnitt* **F** *section efficace de diffusion* **Pl** *przekrój czynny rozpraszania* **Sv** *spridningstvärsnitt*

scattering kernel *rph* • function used in the scattering integral of the transport equation that gives the probability that a particle is scattered and thereby undergoes a certain change in the energy and direction of motion; the *s.k.* is closely related to the differential scattering cross section and it is often denoted $\Sigma_s(E' \rightarrow E, \Omega' \rightarrow \Omega)$, where E', E are the particle energies, and Ω', Ω are the particle vectors of motion before and after the collision, respectively ↦ scattering; transport equation; differential cross section; scattering law **D** *Streufunktion* **F** *noyau de dispersion* **Pl** *jądro rozpraszania* **Sv** *spridningsfunktion; spridningskärna*

scattering law *rph* • scattering kernel with the factor of detailed balance removed; the *s.l.* is commonly denoted $S(\alpha, \beta)$ where α and β depend on the change in the thermal neutron's momentum and energy, respectively ↦ scattering; differential cross section; scattering kernel **D** *Streugesetz* **F** *loi de la diffusion* **Pl** *prawo rozpraszania* **Sv** *spridningslag*

scatter loading *nf* • loading pattern wherein a certain type of fuel (for example, with a certain enrichment or with a certain burnup) is distributed approximately uniformly (but not necessarily regularly) over a given zone of the reactor core ↦ enrichment; burnup; loading pattern **D** *Streubeladung* **F** *chargement réparti* **Pl** *załadunek rozproszony* **Sv** *spridd laddning*

SCC ↦*stress-corrosion cracking*

scintillating material *mt* • material or body that emits light in the form of scintillations, when subject to ionizing radiation ↦ ionizing radi-

ation; scintillation detector **D** *Szintillator* **F** *scintillateur; matériau scintillant* **Pl** *materiał scyntylujący; scyntylator* **Sv** *scintillator*

scintillation detector *mt* • radiation detector comprising a scintillator, which is usually optically coupled to a photosensitive device, e.g., the photomultiplier ↦ radiation detector; scintillator **D** *Szintillationsdetektor* **F** *détecteur à scintillation* **Pl** *detektor scyntylacyjny* **Sv** *scintillationsdetektor*

scintillator ↦*scintillating material*

scram ↦*reactor trip*

scrap *sfg* • (for safeguards:) residues from process steps which contain recyclable nuclear material (unlike waste) ↦ nuclear material; nuclear waste **D** *Schrott* **F** *rebut; résidus* **Pl** *odpad* **Sv** *skrot*

scrubbing (US) *nf* • (for liquid extraction:) removal of pollution or transmission of the pollution to another phase ↦ solvent extraction; extraction cycle **D** *Abscheidung* **F** *lavage* **Pl** *przemywanie* **Sv** *tvätt*

SCWR ↦*supercritical water-cooled reactor*

sealed source *rd* • radiation source in which radioactive material is enclosed in a housing or protected in another way so that there is no risk that the material in normal use is released or otherwise becomes accessible for direct contact ↦ unsealed source; radiation source **D** *umschlossener radioaktiver Strahler; umschlossener radioaktiver Stoff; geschlossenes radioaktives Präparat* **F** *source scellée* **Pl** *zamknięte źródło promieniowania* **Sv** *sluten strålkälla*

secondary coolant *th* • coolant fluid in the secondary coolant circuit ↦ coolant; secondary-coolant circuit **D** *Sekundärkühlmittel* **F** *fluide caloporteur secondaire; caloporteur secondaire; fluide sec-*

ondaire de refroidissement **Pl** *chłodziwo wtórne* **Sv** *sekundärkylmedel*

secondary-coolant circuit *th* ● cooling circuit with which heat is removed from the primary-coolant circuit ↦ primary coolant; primary-coolant circuit; secondary coolant **D** *Sekundärkühlkreislauf; Sekundärkühlkreis* **F** *circuit secondaire de refroidissement* **Pl** *wtórny obieg chłodzenia* **Sv** *sekundärkylkrets*

secondary waste *wst* ● radioactive waste resulting from treatment of the primary waste ↦ radioactive waste; primary waste **D** *Sekundärabfall* **F** *déchets secondaires* **Pl** *odpad wtórny* **Sv** *sekundärt avfall*

seed-core reactor *rty* ● nuclear reactor whose core contains distinct regions with enriched fuel scattered in a lattice of the fuel with lower enrichment or of the fertile material ↦ enriched fuel; fertile material **D** *Saatelementreaktor* **F** *réacteur à coeur à germes* **Pl** *reaktor z mieszanym wzbogaceniem paliwa* **Sv** *blandanrikad reaktor*

seismic sea wave ↦*tsunami*

self-absorption *nap* ● absorption of radiation in the body that emits the radiation ↦ absorption; radiation **D** *Selbstabsorption* **F** *auto-absorption* **Pl** *samopochłanianie; pochłanianie wewnętrzne* **Sv** *självabsorption*

self-powered neutron detector ↦*collectron*

self-regulation *roc* ● property of a nuclear reactor that, at a small deviation from a certain power level, the reactor aims for self-return to the previous level due to the reactivity feedback ↦ reactivity feed-back; reactor stability **D** *Selbstregelung* **F** *autorégulation* **Pl** *samoregulacja* **Sv** *självreglering*

self-shielding *nap* ● shielding of the interior of a body by absorption of the radiation in the outer layer of

the body ↦ shield; absorption; radiation **D** *Selbstabschirmung* **F** *autoprotection* **Pl** *samoosłanianie; samoekranowanie* **Sv** *självskärmning*

semiconductor detector *mt* ● radiation detector whose operation is based on semiconducting properties of a material such as, e.g., silicon or germanium ↦ radiation detector **D** *Halbleiterdetektor* **F** *détecteur semiconducteur* **Pl** *detektor półprzewodnikowy* **Sv** *halvledardetektor*

separation *nf* ● (for isotope separation:) separation factor minus unity ↦ separation factor **D** *Trennung* **F** *coefficient de séparation* **Pl** *separacja* **Sv** *separation²*

separation efficiency *nf* ● ratio of the difference of the isotopic abundance of the inlet and the outlet of a separative element under given operating conditions and the corresponding difference with the same isotopic abundance in the inlet but with the maximum theoretical separation factor ↦ isotopic abundance; separative element; separation factor **D** *Trennwirkungsgrad* **F** *rendement de séparation* **Pl** *wydajność separacji* **Sv** *separationseffektivitet*

separation factor *nf* ● ratio of the isotopic abundance of a particular isotope and the total content of the other isotopes after a separation process, divided by the corresponding ratio before the separation process ↦ isotop; isotopic abundance; isotope separation; separation efficiency **D** *Trennfaktor* **F** *facteur de séparation* **Pl** *współczynnik separacji* **Sv** *separationsfaktor*

separative element *nf* ● minimum separating unit where the elemental isotope separation process occurs ↦ isotope; isotope separation; separative power **D** *Trennglied; Trennstufe* **F** *élément sépara-*

teur **Pl** *element separujący* **Sv** *separerande element*

separative power *nf* • capacity of a separative element ↦ separative element **D** *Trennvermögen* **F** *puissance de séparation* **Pl** *zdolność separacji* **Sv** *separationsförmåga*

separative work *nf* • variable associated with the minimum energy required to separate a certain amount of material with a certain isotope composition into two fractions with different isotopic compositions; the *s.w.* has the dimension of mass and the unit of kg; the *s.w.* can be expressed in terms of the value function $V(x)$; if M_P kg of product with enrichment x_P is obtained from M_F kg of feed with enrichment x_F, the required *s.w.* is given in terms of the *separate work units (SWU)* as $SWU = M_P [V(x_P) - V(x_T)] - M_F [V(x_F) - V(x_T)]$, where x_T is the enrichment of tails ↦ isotope; isotope separation; value function; tails **D** *Trennarbeit* **F** *travail de séparation* **Pl** *praca rozdzielcza* **Sv** *separationsarbete*

separative work unit ↦*separative work*

severe accident ↦*core accident*

SFR ↦*sodium-cooled fast reactor*

shadowing *rph* • local reduction of the particle flux density due to an adjacent absorber; within reactor physics, the *s.* often refers to this reduction of the neutron absorption in an absorber which is caused by a nearby located another absorber ↦ absorber; neutron absorption **D** *Schattenwurf; Abschattung* **F** *effet d'ombre* **Pl** *zacienianie* **Sv** *skuggverkan; skuggning*

shadow shield *rdp* • radiation shield which is placed so that the radiation source is not enclosed but which prevents the passage of radiation in certain directions ↦ shield; radiation **D**

Schattenschild **F** *écran partiel* **Pl** *ekran częściowy* **Sv** *skuggskärm*

shield *rdp* • body with little permeability to radiation intended to reduce the radiation reaching a specific area; the term "radiation protection" should not be used in this sense ↦ radiation; radiation protection; radiation shielding; shielding plug **D** *Abschirmung* **F** *bouclier; écran; blindage* **Pl** *osłona; ekran* **Sv** *strålskärm; skärm*

shielded cell ↦*hot cell*

shielding plug *rdp* • plug adapted to prevent the passage of radiation through a hole in a radiation shield ↦ shield; radiation **D** *Abschirmungsstopfen* **F** *bouchon (dans un blindage)* **Pl** *zatyczka ekranowana* **Sv** *skärmpropp*

shielding window *rdp* • transparent part of the biological shielding ↦ biological shield **D** *Abschirmungsfenster* **F** *hublot* **Pl** *okno ekranowane* **Sv** *skärmfönster*

shift supervisor *roc* • supervisor of a shift of personnel in the control room ↦ control room **D** *Schichtleiter* **F** *chef de quart* **Pl** *kierownik zmiany* **Sv** *skiftingenjör*

shim element *roc* • (also called *shim member:*) control element that is used to provide shimming ↦ control element; shimming **D** *Trimmelement* **F** *élément de compensation* **Pl** *element kompensacyjny; element regulacji zgrubnej* **Sv** *kompensationselement*

shim member ↦*shim element*

shimming *rph* • compensation of slow changes in the reactivity and the neutron flux density in a nuclear reactor ↦ reactivity; neutron flux density **D** *Reaktivitätskompensation* **F** *compensation* **Pl** *kompensacja reaktywności* **Sv** *reaktivitetskompensation*

shipper-receiver difference *sfg* • difference of the quantities of nuclear materials as indicated by the shipping material balance area and the amount measured by the receiving material balance area ↦ nuclear material; material balance area; key measurement point **D** *Absender-Empfänger-Differenz* **F** *différence expéditeur/destinataire* **Pl** *różnica nadajnik-odbiornik* **Sv** *avsändaremottagaredifferens*

shipping cask *rdp* • radiation-shielded vessel for transporting radioactive materials ↦ radiation shielding; radioactive material **D** *Transportbehälter* **F** *conteneur de transport; château de transport* **Pl** *pojemnik transportowy; kontener* **Sv** *transportbehållare*

shipping container ↦*shipping cask*

ship reactor *rty* • nuclear reactor intended for the propulsion of a ship ↦ reactor **D** *Schiffsreaktor* **F** *réacteur de navire* **Pl** *reaktor okrętowy* **Sv** *fartygsreaktor*

shuffling *nf* • relocation of fuel assemblies in order to achieve a more even distribution of the power or burnup in the reactor core ↦ fuel assembly; burnup; reactor core **D** *Brennstoffumsetzung* **F** *réarrangement du combustible* **Pl** *przemieszczanie (paliwa)* **Sv** *bränsleomflyttning*

shutdown[1] *roc* • (of nuclear reactor:) in a clearly subcritical condition ↦ subcritical **D** *abgeschaltet* **F** *arrêté* **Pl** *wyłączony* **Sv** *avställd*

shutdown[2] *roc* • reduction of the reactivity so that the nuclear reactor will be clearly subcritical ↦ reactivity; subcritical **D** *Abschalten* **F** *arrêt* **Pl** *wyłączenie (reaktora)* **Sv** *avställning*

shutdown by control-rod screw insertion *roc* • shutdown of the boiling water reactor by electromechanical screwing of control rods ↦ shutdown[2]; boiling water reactor; control rod **D** *langsame Abschaltung* **F** *arrêt par introduction des barres de commande; arrêt*

normal du réacteur **Pl** *normalne wyłączenie reaktora* **Sv** *skruvstopp*

shutdown for overhaul *roc* • shutdown of a reactor unit for revision and normal refueling ↦ shutdown[2]; reactor unit; refueling **D** *Abschaltung für Revision* **F** - **Pl** *wyłączenie dla przeprowadzenie przeglądu* **Sv** *revisionsavställning*

shutdown margin *roc* • built-in negative reactivity in a nuclear reactor, so that the reactor is clearly subcritical after shutdown ↦ reactivity; subcritical; shutdown[2] **D** *Abschaltreaktivität* **F** *marge d'antiréactivité à l'arrêt* **Pl** *margines reaktywności powyłączeniowej* **Sv** *avställningsmarginal*

shutdown reactivity *rph* • reactivity of a shutdown nuclear reactor ↦ reactivity; shutdown[1] **D** *Abschaltreaktivität* **F** *réactivité à l'arrêt* **Pl** *reaktywność po wyłączeniu reaktora; reaktywność szczątkowa* **Sv** *avställningsreaktivitet*

shutdown to hot stand-by *roc* • shutdown of a reactor unit, wherein the nuclear reactor maintains a temperature near the operating temperature ↦ shutdown[2]; reactor unit **D** *Warmabstellung* **F** *attente à chaud* **Pl** *wyłączenie do stanu gorącego* **Sv** *varmavställning*

shuttle ↦*rabbit*

Si ↦*silicon*

sievert *rdp* • unit of the dose equivalent and the dose commitment, where 1 Sv = 1 J/kg ↦ dose equivalent; dose commitment **D** *Sievert* **F** *sievert* **Pl** *siwert* **Sv** *sievert*

significant quantity *sfg* • amount of nuclear material for which the possibility of producing nuclear weapons can not be excluded ↦ nuclear material **D** *signifikante Stoffmenge* **F** *quantité significative* **Pl** *znaczna ilość* **Sv** *signifikant mängd*

silicon *mat* • chemical element denoted Si, with atomic number $Z=14$, relative atomic mass $A_r=28.0855$, density 2.3296 g/cm^3, melting point 1414 °C, boiling point 3265 °C, crustal average abundance 2.82×10^5 mg/kg and ocean abundance 2.2 mg/L ↦ element **D** *Silicium* **F** *silicium* **Pl** *krzem* **Sv** *kisel*

silver *mat* • chemical element denoted Ag, with atomic number $Z=47$, relative atomic mass $A_r=107.8682$, density 10.501 g/cm^3, melting point 961.78 °C, boiling point 2162 °C, crustal average abundance 0.075 mg/kg and ocean abundance 4×10^{-5} mg/L; one of the metals used for manufacturing of control rods ↦ control rod **D** *Silber* **F** *argent* **Pl** *srebro* **Sv** *silver*

simple cascade *nf* • cascade in which the enriched fraction is passed to the next succeeding stage and the depleted fraction to the previous one ↦ cascade; enriched material **D** *einfache Kaskade* **F** *cascade simple* **Pl** *kaskada prosta* **Sv** *enkel kaskad*

single scattering *nap* • deflection of a particle from the initial path caused by a collision with a single scattering center ↦ scattering **D** *Einfachstreuung* **F** *diffusion simple* **Pl** *rozproszenie jednokrotne; rozproszenie pojedyncze* **Sv** *enkelspridning*

site area *rs* • geographical area that contains an authorized facility, authorized activity or source, and within which the management of the authorized facility or authorized activity may directly initiate emergency actions; the *s.a.* is typically the area within the security perimeter fence or other designated property marker ↦ controlled area; supervised area **D** - **F** - **Pl** *teren lokalizacji* **Sv** -

site area emergency ↦reference level for emergency action

siting criterion *rs* • criterion for assessing how suitable a site for a nuclear installation is, especially considering nuclear safety ↦ site area; nuclear installation; nuclear safety **D** *Standortkriterium* **F** *critère de choix d'un site* **Pl** *warunek (wyboru) lokalizacji* **Sv** *förläggningskriterium*

skyshine *rd* • ionizing radiation reaching an object that is spread by the air; the *s.* commonly refers to the radiation scattered above a radiation shielding wall, but sometimes to the radiation scattered from surrounding objects ↦ ionizing radiation **D** *Luftstreuung* **F** *effet de ciel* **Pl** *promieniowanie rozproszone w powietrzu* **Sv** *luftspridd strålning*

slowing-down area *rph* • one-sixth of the mean square of the distance travelled by a neutron from a point source in an infinite, homogeneous medium, to the point where the neutron decelerates from the source energy to any specific energy ↦ slowing-down length; slowing-down kernel; slowing-down density; slowing-down power **D** *Bremsfläche* **F** *aire de ralentissement* **Pl** *powierzchnia spowalniania* **Sv** *bromsarea*

slowing-down density *rph* • (denoted q:) rate at which slowing-down neutrons pass through a certain energy level per unit volume and time, given as: $q = \xi\Sigma_s\phi(\mathbf{r}, u)$, where ξ - the average logarithmic energy decrement per collision, Σ_s - the macroscopic cross section for scattering and $\phi(\mathbf{r}, u)$ - the neutron flux density per unit lethargy u at the space location $\mathbf{r}$ ↦ average logarithmic energy decrement; neutron flux density; lethargy **D** *Bremsdichte* **F** *densité de ralentissement* **Pl** *gęstość spowalniania* **Sv** *bromstäthet*

slowing-down kernel *rph* • function which describes the probability that a neutron in a homogeneous substance moves from a certain location to another one while it slows down by a certain energy interval ↦ slowing-down density; slowing-down area **D** Bremskern **F** noyau de l'intégrale de ralentissement **Pl** jądro spowolniania **Sv** bromskärna

slowing-down length rph • length calculated as the square root of the slowing-down area ↦ slowing-down area **D** Bremslänge **F** longueur de ralentissement **Pl** długość spowalniania **Sv** bromslängd

slowing-down power rph • measure of the ability of a material to slow down neutrons; the s.-d.p. is defined as the product of the average logarithmic energy decrement of the material and its macroscopic scattering cross section ↦ slowing-down area; average logarithmic energy decrement; macroscopic cross section; scattering cross section **D** Neutronenbremsvermögen **F** pouvoir de ralentissement **Pl** zdolność spowalniania; zdolność moderacji **Sv** neutronbromsförmåga

slow neutron rd • neutron with energy lower than an upper limit that in reactor physics is often set at 1 eV ↦ neutron; reactor physics; epithermal neutron; fast neutron **D** langsames Neutron **F** neutron lent **Pl** neutron powolny **Sv** långsam neutron

slug flow th • two-phase flow pattern characterized by a series of liquid slugs (plugs), separated by relatively large gas pockets; in vertical flow, the gas pockets have generally blunt heads, flat ends, and occupy almost the entire cross-sectional area of the tubing ↦ two-phase flow; bubbly flow; annular flow; mist flow **D** - **F** écoulement à poches **Pl** przepływ korkowy **Sv** ångkuddeflöde

slurry reactor rty • (also called suspension reactor:) dispersion reactor in

which a solid fuel is dispersed in a liquid ↦ reactor **D** Suspensionsreaktor **F** réacteur à combustible en suspension **Pl** reaktor zawiesinowy; reaktor na paliwo zawiesinowe **Sv** suspensionsreaktor

Sm ↦samarium

smear test rdp • investigation of possible radioactive contamination on surfaces, e.g., at workplaces or on sources; the s.t. is performed by a patch (or similar) of a porous material that is drawn across the surface to be checked, then the patch activity is measured ↦ radioactive contamination; radioactivity **D** Wischtest **F** frottis **Pl** pobieranie próbki przez pocieranie **Sv** strykprovtagning

SMR rty • (acronym for Small Modular Reactor:) reactor concept with intentionally reduced power output, typically less than 300 MWe, designed for niche applications in remote or isolated areas; main features of the s.m.r. include: (i) modular design (ii) high degree of innovation (iii) addressing conditions and requirements of target markets ↦ pressurized water reactor **D** SMR **F** SMR **Pl** SMR **Sv** SMR

SNAP th • (acronym for Symbolic Nuclear Analysis Package:) graphical user interface with pre-processor and post-processor capabilities, which assists users in developing TRACE and RELAP5 input decks and running the codes ↦ RELAP; TRACE **D** SNAP **F** SNAP **Pl** SNAP **Sv** SNAP

sodium mat • chemical element denoted Na, with atomic number $Z=11$, relative atomic mass $A_r=22.98977$, density 0.97 g/cm^3, melting point 97.8 °C, boiling point 883 °C, crustal average abundance 2.36×10^4 mg/kg and ocean abundance 1.08×10^4 mg/L; the liquid s.

is used as the coolant in the sodium-cooled fast reactor ↦ element; sodium-cooled fast reactor **D** Natrium **F** sodium **Pl** sód **Sv** natrium

sodium-cooled fast reactor *rty* • (abbreviated *SFR:*) generation IV nuclear reactor with a fast neutron spectrum whose core is cooled with liquid sodium ↦ generation IV reactor; sodium **D** schneller natriumgekühlter Reaktor **F** réacteur rapide refroidi au sodium **Pl** reaktor prędki chłodzony sodem **Sv** natriumkyld snabb reaktor

sodium hot-trap *rcs* • apparatus for removal of contaminants, usually sodium oxide, from circulating sodium, by contact with a solid with which the contaminants react at an elevated temperature ↦ sodium; cold trap **D** Heißfalle für Natrium **F** piège chaud **Pl** pułapka sodowa gorąca **Sv** hetfälla

solidification *wst* • transformation of liquids or liquid-like substances, e.g., radioactive waste, into a solid ↦ radioactive waste **D** Verfestigung **F** solidification **Pl** krzepnięcie; zestalanie się **Sv** solidifiering

solvent extraction *nf* • process in which a substance is selectively extracted from one phase into another one in a system of two immiscible phases, one of which is usually in a liquid phase ↦ stripping **D** Flüssigkeitsextraktion; Lösungsmittelextraktion **F** extraction par solvant **Pl** ekstrakcja rozpuszczalnikowa; ekstrakcja w układzie ciecz-ciecz **Sv** vätskeextraktion

somatic effect of radiation *nf* • radiation effect on the somatic cells, that is, those cells which are not germ cells; somatic radiation effect can lead to, e.g., radiation sickness or the occurrence of cancer ↦ radiation sickness **D** somatische Strahlenwirkung **F** effet somatique des rayonnements **Pl** somatyczny

skutek promieniowania **Sv** somatisk strålningsverkan

source data *sfg* • data used to identify a nuclear material ↦ nuclear material **D** Primärdaten **F** données de base **Pl** dane źródłowe **Sv** källdata

source material *sfg* • natural uranium, depleted uranium or thorium, and materials containing these substances in an amount that can have nuclear use; the term *s. m.* has been previously used in legal texts ↦ natural uranium; depleted uranium; thorium; significant quantity **D** Ausgangsstoff **F** matières brutes **Pl** materiał źródłowy **Sv** atområbränsle

source range *roc* • power range of a nuclear reactor in which an extra neutron source is required to allow measurement of the neutron flux density ↦ reactor; neutron source; neutron flux density; power range **D** Quellenbereich **F** niveau des sources; niveau des chaînes-sources; domaine des sources **Pl** zakres pracy reaktora ze źródłem pomocniczym **Sv** källområde

source range monitor *roc* • monitor of reactor power in the source-range region ↦ power-range monitor; source range **D** Quellenbereichsmonitor **F** chaîne-source; moniteur du domaine des sources **Pl** monitor zakresu pracy reaktora ze źródłem pomocniczym **Sv** källeffektkanal

source reactor *rty* • nuclear reactor intended for use as a neutron source for experiments ↦ reactor; power reactor; research reactor **D** Meßreaktor; Neutronenquellreaktor **F** réacteur source **Pl** reaktorowe źródło neutronów **Sv** källreaktor

source term ↦radioactive source term

spacer *nf* • construction element of a fuel bundle which supports fuel rods and prevents their vibration, bending and lateral displacements ↦ fuel bundle; fuel rod **D** Abstandshalter **F** espaceur;

pièce d'espacement **Pl** *element dystansujący* **Sv** *spridare*

spacer capture rod *nf* • (in fuel bundle:) rod device which limits the vertical movement of the spacer ↦ fuel bundle; spacer **D** *Brennstab mit Vorrichtung für Abstandshalter* **F** *aiguille (de combustible) avec dispositif d'espaceur* **Pl** *pręt paliwowy z blokadą dla elementu dystansującego* **Sv** *spridarhållarstav*

spacer grid *nf* • (in fuel assembly:) spacer with a grid-like design ↦ spacer; fuel assembly **D** *Halterung für die Brennstäbe* **F** *grille d'espacement* **Pl** *siatka dystansująca* **Sv** *spridargaller*

spallation *nap* • process in which fragments of material are ejected from a body due to impact or stress; nuclear *s.* is one of the processes by which a particle accelerator may be used to produce a beam of neutrons; a particle beam consisting of protons at around 1 GeV is shot into a target consisting of mercury, tantalum, lead or another heavy metal; consequently, the target nuclei are excited and upon de-excitation, 20 to 30 neutrons are expelled per a nucleus; energetic cost of one *s.* neutron is six times lower than that of a neutron gained via nuclear fission; in contrast to nuclear fission, the *s.* neutrons cannot trigger further *s.* or fission processes to produce further neutrons; therefore, there is no chain-reaction, which makes the process non-critical ↦ accelerator-driven system **D** *Spallation* **F** *spallation* **Pl** *spalacja* **Sv** *spallation*

special nuclear material *sfg* • enriched uranium or plutonium, and materials containing these substances; the term is used, among others, in international documents ↦ nuclear material; enriched uranium; plutonium **D** *besonderer spaltbarer Stoff* **F** *matières*

nucléaires spéciales **Pl** *specjalny materiał jądrowy* **Sv** *särskilt klyvbart material*

specific activity *rdy* • (for given radioactive material:) ratio of the activity of a radioactive sample to the mass of the sample expressed in, e.g., Bq/g; for a pure sample the *s.a.* $\hat{A}$ can be calculated as $\hat{A} \equiv$ activity/mass $= \lambda N_A/M$, where λ - radionuclide decay constant, N_A - Avogadro's constant, M - radionuclide molecular mass ↦ radioactive material; activity; becquerel; Avogadro's constant; molecular mass **D** *spezifische Aktivität* **F** *activité massique* **Pl** *aktywność właściwa* **Sv** *specifik aktivitet*

specific burnup *rph* • total energy developed per unit mass of nuclear fuel, often expressed in megawatt-days per kilogram of heavy metal ($MWd/kgHM$); for clarity the nuclear fuel needs to be specified ↦ nuclear fuel; burnup **D** *spezifischer Abbrand* **F** *épuisement spécifique* **Pl** *głębokość wypalenia* **Sv** *specifik utbränning*

specific gamma-ray constant *rdp* • constant that is calculated as the exposure rate (exposure per unit time) at a certain distance from a point-radiation source, containing a particular radioactive nuclide that emits gamma radiation, multiplied by the square of the distance divided by the radiation source activity, when neglecting attenuation ↦ radiation source; nuclide; gamma radiation; attenuation **D** *spezifische Gammastrahlenkonstante* **F** *constante spécifique de rayonnement gamma* **Pl** *stała właściwa promieniowania gamma* **Sv** *expositionsratkonstant*

specific gas constant *th* • constant of a gas or a mixture of gases given by the universal gas constant divided by the molar mass of the gas or the gas mixture ↦ universal gas constant; molar mass **D** *spezifische Gaskonstante* **F** *con-*

stante spécifique d'un gaz **Pl** indywidualna stała gazowa **Sv** individuella gaskonstanten
specific power *th* • power output per mass of the fuel in a nuclear reactor ⟼ power density   **D** spezifische Leistung **F** puissance massique **Pl** moc właściwa **Sv** specifik effekt
spectral shift control *roc* • particular form of a moderator control of a nuclear reactor, in which the neutron energy spectrum is intentionally modified ⟼ reactor control[2]; moderator control; neutron flux density   **D** Spektralsteuerung **F** commande par réglage du spectre; commande par dérive spectrale **Pl** sterowanie (reaktora) przesunięciem widma **Sv** spektrumstyrning
spectral shift reactor *rty* • reactor whose neutron spectrum can be changed, for control or other purposes, by a change of the moderator amount or its properties ⟼ spectral shift control; neutron flux density   **D** Reaktor mit Spektralsteuerung **F** réacteur à dérive spectrale; réacteur à spectre réglable **Pl** reaktor o zmiennym widmie **Sv** spektralskiftsreaktor
spectroscopy *bph* • branch of physics and chemistry concerned with investigation of structures and properties of molecules, atoms and nuclei ⟼ atom; nucleus   **D** Spektroskopie **F** spectroscopie **Pl** spektroskopia **Sv** spektroskopi
spent fuel *nf* • nuclear fuel that has been removed from a nuclear reactor after having undergone burnup and that should no longer be used in this reactor ⟼ burnup   **D** verbrauchter Brennstoff **F** combustible usé **Pl** paliwo wypalone **Sv** utbränt bränsle; använt bränsle
spent fuel pool *rcs* • water-filled pool where fuel assemblies can be handled, e.g., in connection with refuelling or temporary storage ⟼ fuel handling; refuelling; interim storage   **D** Bren-

nelementbecken **F** piscine de stockage du combustible usé **Pl** basen wypalonego paliwa **Sv** bränslebassäng
spherical harmonics method *rph* • approximate solution method of the transport equation based on the expansion of the angular dependence of the particle flux density with the spherical harmonics; when the expansion series ends after $L + 1$ terms, the so-called P_L approximation is obtained ⟼ particle flux density   **D** Kugelfunktionsmethode **F** méthode des harmoniques sphériques **Pl** metoda funkcji kulistych **Sv** klotfunktionsmetod
spike *mat* • (in materials technology:) area of a crystal disordered by a particle with a high linear energy transfer ⟼ linear energy transfer   **D** Störzone **F** zone des dégâts **Pl** strefa uszkodzenia (popromiennego) **Sv** strålskadezon
SPN detector ⟼collectron
spontaneous fission *nap* • fission which takes place without any energy supplied to the nucleus by particles or by other means ⟼ fission   **D** spontane Spaltung **F** fission spontanée **Pl** rozszczepienie spontaniczne **Sv** spontan fission
spray cooling *th* • cooling by water spraying of, e.g., a reactor core at the emergency core cooling, or a reactor containment in a loss-of-coolant accident, when steam has discharged from the primary-coolant circuit ⟼ reactor core; containment[1]; emergency core cooling; loss-of-coolant accident; primary-coolant circuit   **D** Spritzkühlung **F** refroidissement par aspersion **Pl** chłodzenie natryskowe **Sv** strilkylning
square cascade *nf* • (at isotope separation:) cascade in which the mass flow is the same in each stage ⟼ cascade; stage; isotope separation; separative element; separative work   **D** Stufenkaskade **F**

cascade constante **Pl** *kaskada równomierna* **Sv** *fyrkantskaskad*

SSE ↦*safe shutdown earthquake*

stable nuclide *rdy* • non-radioactive nuclide ↦ nuclide; unstable nuclide **D** *stabiles Nuklid* **F** *nucléide stable* **Pl** *nuklid trwały* **Sv** *stabil nuklid*

stable reactor period ↦*reactor time constant*

stage *nf* • (at isotope separation:) part of the cascade in which all devices are working in parallel on materials with the same isotopic composition ↦ isotope separation; cascade **D** *Stufe* **F** *étage* **Pl** *stopień* **Sv** *steg*

standard tails assay *nf* • (at isotope separation:) adopted residue of a cascade when evaluating the isotope separation plant operations and economy ↦ isotope separation; cascade **D** *Nennabstreifkonzentration* **F** *teneur de rejet normalisée* **Pl** *nominalny ogon separacji* **Sv** *nominell resthalt*

Stanton number *th* • (denoted St:) dimensionless number that represents the ratio of heat transferred into a fluid to the thermal capacity of the fluid, defined as, $St = Nu/RePr = \frac{h}{\rho U c_p}$, where Nu - Nusselt number, Re - Reynolds number, Pr - Prandtl number, h- heat transfer coefficient, ρ - fluid density, U - fluid speed, and c_p - fluid specific heat at constant pressure ↦ Nusselt number; Reynolds number; Prandtl number **D** *Stanton-Zhal* **F** *nombre de Stanton* **Pl** *liczba Stantona* **Sv** *Stantonstal*

start-up ↦*commissioning*

start-up neutron source *roc* • neutron source placed in a subcritical nuclear reactor to increase the neutron flux density and thus facilitate monitoring at the upturn to criticality ↦ neutron source; criticality **D** *Anfahrneutronenquelle* **F** *source (de neutrons) de démar-*

rage **Pl** *rozruchowe źródło neutronów* **Sv** *startneutronkälla*

static quality ↦*steam quality*

station blackout *rs* • (abbreviated *SBO*:) internal event routinely considered in safety analysis of nuclear power plants, consisting of a power failure in combination with a failed transition to the house load operation and loss of the diesel engine backup facility ↦ event; house load operation **D** *Kraftwerksausfall* **F** *perte totale des alimentations électriques; perte d'alimentation générale* **Pl** *całkowita utrata zasilania elektrycznego* **Sv** *totalt elbortfall*

steam dryer *rcs* • (in boiling water reactor:) apparatus located within the reactor vessel above the reactor core to separate the residual water from steam ↦ boiling water reactor; reactor vessel; reactor core **D** *Dampftrockner* **F** *séparateur d'humidité[2]* **Pl** *osuszacz pary* **Sv** *fuktavskiljare[2]; ångtork*

steam explosion *th* • explosion caused by violent boiling or flashing of water into steam, occurring when the water is either superheated, rapidly heated by fine hot debris produced within it, or heated by the interaction of molten metals as in, e.g., fuel–coolant interactions of molten nuclear-reactor fuel rods with water in a nuclear reactor core, following the core meltdown; during *s.e.* the water changes from the liquid phase to the gas phase with extreme speed and with a dramatic increase in volume ↦ boiling; flashing; core meltdown **D** - **F** *explosion de vapeur d'eau* **Pl** *eksplozja pary wodnej* **Sv** *ångsexplosion*

steam generator *rcs* • heat exchanger included in the nuclear steam-supply system, in which water is turned into steam ↦ nuclear steam-supply system **D** *Dampferzeuger* **F**

générateur de vapeur **Pl** *wytwornica pary* **Sv** *ånggenerator*

steam quality *th* • (also called *quality:*) measure of the vapour mass content in the two-phase mixture; the *static quality* x_s is defined as a ratio of the vapour mass m_v in a mixture to the total mass of the mixture $m = m_v + m_l$, where m_l is the mass of the liquid phase, thus $x_s = m_v/m$; the *flow quality* (also called the *actual quality*) x is defined as a ratio of the vapour mass flow rate W_v to the total mass flow rate of the mixture $W = W_v + W_l$, where W_l is the mass flow rate of the liquid phase, thus $x = W_v/W$; the *thermodynamic equilibrium quality* (also called the *mixing-cup quality*) is defined as $x_e = (i_m - i_f)/i_{fg}$, where x_m is the specific enthalpy of the two-phase mixture, i_f is the specific enthalpy of the saturated liquid and i_{fg} is the heat of vaporization ↦ two-phase flow; mass flow rate; heat of vaporization **D** *Dampfqualität* **F** *titre en vapeur* **Pl** *stopień suchości pary* **Sv** *ångkvalitet*

steam separator *rcs* • (in a boiling water reactor:) device placed above the reactor core with the task to separate steam from the steam-water mixture that is rising from the reactor core ↦ boiling water reactor **D** *Dampfabscheider* **F** *séparateur de vapeur* **Pl** *separator pary* **Sv** *ångavskiljare; ångseparator*

Stefan-Boltzmann   constant ↦*Stefan-Boltzmann law*

Stefan-Boltzmann law *bph* • physical law that describes the power radiated from a black body in terms of its absolute temperature and states that the total energy radiated per unit surface area of the black body across all wavelengths per unit time is directly proportional to the fourth power of

the black body's absolute temperature T: $q" = \sigma T^4$, where σ is the *Stefan-Boltzmann constant* given as $\sigma = 2\pi^5 k_B^4/(15c^2h^3) = 5.670\,373 \times 10^{-8}$ W/(m²K⁴), in which k_B is the Boltzmann constant, h is Planck's constant and c is the speed of light in vacuum ↦ absolute temperature; Boltzmann constant; Planck's constant; Wien's displacement law **D** *Stefan-Boltzmann-Gesetz* **F** *loi de Stefan-Boltzmann; loi de Stefan* **Pl** *prawo Stefana-Boltzmanna; prawo Stefana* **Sv** *Stefan-Boltzmanns lag*

step insertion of reactivity *rph* • sudden change of the reactivity in a critical reactor at a specific time t (usually $t = 0$ is assumed) to a non-zero value ρ_0, which can be either positive or negative ↦ reactivity; ramp insertion of reactivity **D** *stufenförmige Reaktivitätsänderung* **F** *apport de réactivité par paliers* **Pl** *skokowe wprowadzenie reaktywności* **Sv** *reaktivitetssteg*

stored energy *mat* • internal energy supplied to a solid when exposed to ionizing radiation ↦ ionizing radiation **D** *gespeicherte Energie* **F** *énergie emmagasinée* **Pl** *energia zakumulowana* **Sv** *lagrad energi*

strain ↦*strength of materials*

strategic point *sfg* • place in a nuclear facility at which the necessary and sufficient information on safeguards can be obtained ↦ safeguards (of nuclear materials) **D** *strategischer Punkt* **F** *point stratégique* **Pl** *punkt strategiczny* **Sv** *strategisk punkt*

streaming ↦*channeling effect*

strength of materials *mat* • engineering discipline concerned with the ability of a material to resist mechanical forces when in use; the material's strength in a given application depends on many factors, including its resistance to deformation and cracking, and it often depends on the shape of the member being designed; a metallic solid will suffer a deformation (or *strain*) when subjected to a load (or *stress*); for relatively low stresses, the metal exhibits elastic behaviour and it will recover its original configuration when the load is removed; for high enough strains the metal is deformed in a plastic manner, and it will not recover the initial shape; the stress corresponding to the transition from elastic to plastic behaviour is called the *yield stress* (or *yield strength*); for the elastic region the relationship between stress σ (defined here as a load in newtons per unit area in square meters) and strain ϵ (defined here as a dimensionless ratio of the increase in dimension to the initial dimension) is given by the following *Hook's law*: $\sigma = E\epsilon$, where E is *Young's modulus of elasticity*, which is a material property and is roughly 2×10^{11} Pa for most steels ↦ stress-corrosion cracking **D** *Festigkeitslehre* **F** *résistance des matériaux* **Pl** *wytrzymałość materiałów* **Sv** *hållfasthetslära*

stress ↦*strength of materials*

stress corrosion *mat* ● process that occurs by interaction between corrosion and strains in metal due to tension ↦ stress-corrosion cracking; strength of materials **D** *Spannungskorrosion* **F** *corrosion sous contrainte* **Pl** *korozja naprężeniowa* **Sv** *spänningskorrosion*

stress-corrosion cracking *mat* ● cracking due to stress corrosion ↦ stress corrosion **D** *Spannungsrißkorrosion* **F** *fissuration sous contrainte* **Pl** *pękanie wskutek korozji naprężeniowej* **Sv** *spänningskorrosionssprickning*

stretch-out *roc* ● extension of a reactor unit's operating period, in addition to the reactor core's rated life at full power, by changing the operating conditions, e.g., by increasing the coolant flow rate in a boiling water reactor, or reducing the primary fluid temperature in a pressurized water reactor, in order to offset the decline in reactivity caused by the fuel burnup ↦ reactor unit; operating period **D** *Betriebsverlängerung* **F** *exploitation en allongement de cycle* **Pl** *wydłużenie kampanii paliwowej (reaktora)* **Sv** *driftförlängning*

stripping (GB) ↦*scrubbing (US)*

stripping (US) *nf* ● (also called *backwash(GB):*) process in which a substance removed by solvent extraction is reversed from the organic phase to the water solution ↦ solvent extraction **D** *Rückextraktion* **F** *réextraction* **Pl** *usuwanie powłoki* **Sv** *återextraktion*

strontium *mat* ● chemical element denoted Sr, with atomic number Z=38, relative atomic mass A_r=87.62(1), density 2.64 g/cm^3, melting point 777 °C, boiling point 1377 °C, crustal average abundance 0.36 mg/kg and ocean abundance 8 mg/L ↦ element; nuclear battery **D** *Strontium* **F** *strontium* **Pl** *stront* **Sv** *strontium*

subchannel *th* ● hypothetical part of a flow channel containing physical and imaginary walls in which the thermal-hydraulic aspects of the flow can be investigated ↦ subchannel analysis **D** *Unterkanal* **F** *sous-canaux* **Pl** *podkanał* **Sv** *delkanal*

subchannel analysis *th* ● (in thermal-hydraulic reactor calculations:) thermal-hydraulic model consisting of a number of parallel subchannels, for which separate sets of mass, momentum and energy conservation equations are formulated, with particular regard to the coupling between subchannels ↦ subchannel **D**

Unterkanalanalyse **F** *analyse de sous-canaux* **Pl** *analiza podkanałowa* **Sv** *delkanalanalys*

subcooled *th* ● condition when the system temperature is below the saturation temperature at the prevailing local pressure ↦ subcooled boiling; subcooled film boiling; subcooled nucleate boiling **D** *unterkühlt,-es* **F** *sous-saturé,-e* **Pl** *przechłodzony,-e,-a* **Sv** *underkyld*

subcooled boiling *th* ● boiling heat transfer when the fluid has reached the boiling point near the heated surface but its bulk is still subcooled and vapour bubbles exist only near the heated wall surface ↦ heat transfer; subcooled; subcooled film boiling; saturated boiling **D** *unterkühltes Sieden* **F** *ébullition sous-saturée; ébullition locale* **Pl** *wrzenie przechłodzone* **Sv** *underkyld kokning*

subcooled film boiling *th* ● film boiling in which the wall vapour layer is created when the bulk liquid phase is subcooled ↦ film boiling; subcooled; subcooled nucleate boiling **D - F - Pl** *wrzenie błonowe przechłodzone* **Sv** -

subcooled nucleate boiling *th* ● nucleate boiling in which vapour bubbles are formed at the heated wall surface when the bulk liquid is still subcooled ↦ nucleate boiling; subcooled **D - F - Pl** *wrzenie pęcherzykowe przechłodzone* **Sv** *underkyld bubbelkokning*

subcritical *rph* ● not capable of maintaining an independent chain reaction; this means that the system in which the chain reaction occurs has an effective multiplication factor less than unity ↦ critical; supercritical **D** *unterkritisch* **F** *sous-critique* **Pl** *podkrytyczny* **Sv** *underkritisk*

subcritical assembly *rph* ● device containing a multiplying or non-multiplying medium and configured to be subcritical; the *s.a.* is usually used together with an independent

source of neutrons to determine the properties of the medium when interacting with neutrons ↦ multiplying; neutron source **D** *unterkritische Anordnung* **F** *assemblage sous-critique* **Pl** *zestaw podkrytyczny* **Sv** *underkritisk uppställning*

subcritical multiplication *rph* ● ratio of the total number of neutrons, originating from fission and from a constant source, which at equilibrium exist in a subcritical system, and the total number of neutrons that would exist in the system in the absence of fission ↦ fission; subcritical **D** *unterkritische Multiplikation* **F** *multiplication sous-critique* **Pl** *mnożenie podkrytyczne* **Sv** *underkritisk multiplikation*

sun-burst *mat* ● attack of hydrogen-containing substances in the zircaloy cladding of a fuel rod, which leads to a local precipitation of zirconium ↦ zircaloy; cladding; zirconium **D** *lokaler Hydridangriff* **F** *hydruration* **Pl** *lokalne wytrącenie cyrkonu spowodowane wodorkami* **Sv** *lokalt hydridangrepp*

supercell *rph* ● portion of a reactor lattice containing multiple elements that are characteristic of a reactor core, e.g., a control rod with surrounding fuel assemblies ↦ reactor lattice; control rod; fuel assembly **D** *Superzelle* **F** *supercellule* **Pl** *podzespół rdzenia* **Sv** *supercell*

supercritical *rph* ● able to get a chain reaction to grow, i.e., a system in which the chain reaction occurs has an effective multiplication factor greater than unity ↦ critical; subcritical **D** *überkritisch* **F** *surcritique* **Pl** *nadkrytyczny* **Sv** *överkritisk*

supercritical fluid *th* ● fluid at a temperature and pressure above its critical point values, where distinct liquid and gas phases do not exist; in particular *supercritical steam* exists when ordinary water is under pressure exceeding the critical pressure $p_c = 22.06$ MPa; the area on the pT-phase diagram for water, where both pressure p and temperature T exceed their corresponding critical values (that is, $p > 22.06$ MPa and $T > 647$ K), is called the *supercritical zone* ↦ critical point **D** - **F** - **Pl** *płyn w stanie nadkrytycznym* **Sv** *superkritisk vätska*

supercritical steam ↦*supercritical fluid*

supercritical water-cooled reactor *rty* • (abbreviated SCWR:) generation IV reactor concept, designed as (primarily) light-water reactor that uses the supercritical steam as the coolant and moderator ↦ generation IV reactor; supercritical fluid; light-water reactor **D** *überkritischer Leichtwasserreaktor* **F** *réacteur à eau supercritique* **Pl** *reaktor chłodzony wodą w stanie nadkrytycznym* **Sv** *superkritiskvattenkyld reaktor*

supercritical zone ↦*supercritical fluid*

superficial individual dose equivalent *rdp* • dose equivalent in a tissue (with density 1000 kg/m^3) at the depth of 0.07 mm inside the body; the quantity is intended for personal dosimetry and can be related to the effective dose equivalent; the unit for *s.i.d.e.* is the sievert ↦ dose equivalent; penetrating individual dose equivalent; effective dose equivalent; sievert **D** - **F** *équivalent superficiel de dose individuel* **Pl** *powierzchniowy równoważnik dawki* **Sv** *ytlig individdosekvivalent*

superficial velocity *th* • hypothetical flow velocity calculated as if the given phase or fluid were the only one flowing or present in a given cross-sectional area; for phase k of a multiphase mixture flowing with mass flux G_k, volumetric flow rate Q_k, and

cross-section averaged velocity U_k, in a channel with cross-section area A, the *s.v.* J_k can be found as $J_k = Q_k/A = G_k/\rho_k = \alpha_k U_k$, where ρ_k and α_k are the mass density and the volume fraction of the phase k, respectively ↦ two-phase flow **D** - **F** *vitesse apparente; vitesse superficielle* **Pl** *prędkość pozorna* **Sv** *skenbar hastighet*

superheat reactor *rty* • nuclear reactor where coolant is superheated inside or outside of the reactor core using the reactor's heat ↦ nuclear superheat; core **D** *Überhitzungsreaktor* **F** *réacteur à surchauffe* **Pl** *reaktor z przegrzewem jądrowym* **Sv** *överhettarreaktor*

supervised area *rs* • defined area, not designated as a controlled area, but for which occupational exposure conditions are kept under review, even though no specific protection measures or safety provisions are normally needed ↦ controlled area; site area **D** - **F** - **Pl** *obszar nadzorowany* **Sv** -

supporting fuel rod *nf* • rod in a fuel bundle, which constitutes a fixed connection between the top tie plate and the bottom tie plate (in BWRs) or between the top nozzle and the bottom nozzle (in PWRs) ↦ fuel rod; fuel bundle; top tie plate; bottom tie plate; top nozzle; bottom nozzle **D** *tragender Brennstab* **F** *barre de combustible support* **Pl** *nośny pręt paliwowy* **Sv** *bärande bränslestav*

surface heat flux *th* • (usually denoted q'':) transferred heat per unit time and area of a fuel rod or a fuel element; for a fuel rod with the cladding outer diameter d_{Co} and the linear power density q', the *s.h.f.* is given as $q'' = q'/(\pi d_{Co})$ ↦ heat flux; fuel rod; fuel element; cladding; linear power density **D** *Heizflächenleistungsdichte* **F** *puissance sur-*

facique **Pl** *powierzchniowy strumień ciepła*
Sv *yteffekt*
surface power density ↦*surface heat*
flux
suspension reactor ↦*slurry reactor*
swelling *mat* • volume increase of the
fuel element due to, among others, irradiation ↦ fuel element; irradiation **D**
Schwellen **F** *gonflement* **Pl** *puchnięcie* **Sv**
svällning

swimming-pool reactor ↦*pool reactor*
SWU ↦*separative work*
synthetic element *bph* • chemical element not existing in nature,
such as, e.g., transuranic elements ↦
transuranic element **D** - **F** *élément synthétique* **Pl** *pierwiastek sztuczny; pierwiastek
syntetyczny* **Sv** *syntetiskt grundämne; konstgjort grundämne*

Tt

T ↦ *tritium*

tail end *nf* • (at fuel reprocessing:) final conversion of spent fuel in order to give the final form to purified products ↦ fuel reprocessing; spent fuel; headend **D** *Endreinigung* **F** *traitement final* **Pl** *obróbka końcowa* **Sv** *slutsteg*[1]

tails *nf* • (at isotope separation:) material leaving a cascade, depleted of the desired isotope ↦ isotope separation; cascade; depleted material; isotope **D** *Abfall* **F** *rejet* **Pl** *frakcja końcowa* **Sv** *restfraktion*

tank reactor *nf* • heterogeneous reactor with the reactor core contained in a vessel ↦ heterogeneous reactor; reactor vessel **D** *Tankreaktor* **F** *réacteur à coeur fermé* **Pl** *reaktor zbiornikowy* **Sv** *tankreaktor*

target *nf* • material or object which is irradiated or bombarded ↦ irradiation **D** *Target* **F** *cible* **Pl** *tarcza* **Sv** *strålmål*

technical radiology ↦ *radiology*

technical specification for reactor operation ↦ *operating rules*

temperature coefficient of reactivity *rph* • (for reactivity of nuclear reactor:) reactivity coefficient with respect to the temperature of a particular material or in a certain point in a nuclear reactor; the *t.c.o.r.* is usually divided into prompt and delayed components, resulting from the instantaneous state of the fuel due to the Doppler effect or thermal distortion of fuel elements, and the thermal expansion of the moderator material, respectively ↦ reactivity; reactivity coefficient; Doppler effect; Doppler coefficient **D** *Temperaturkoeffizient (der Reaktivität)* **F** *coefficient de température (de la réactivité)* **Pl** *temperaturowy współczynnik reaktywności* **Sv** *temperaturkoefficient*

temporary absorber *rph* • neutron absorber present in a nuclear reactor over a limited period of operation when the excess reactivity or neutron flux differ from the normal values, e.g., during the running-in period; the absorbing effect of a *t.a.* does not usually change significantly over the operating period ↦ neutron absorber; excess reactivity; neutron flux density; running-in period; operating period **D** *temporärer Absorber* **F** *absorbeur temporaire* **Pl** *tymczasowy absorbent neutronów* **Sv** *tillfällig absorbator*

ternary fission *nap* • rarely occurring nuclear fission where three fission fragments are formed; one of the fission fragments can be a light nucleus ↦ nuclear fission; fission fragment **D** *Kernspaltung in drei Kernbruchstücke* **F** *tripartition; fission ternaire* **Pl** *rozszczepienie trójfragmentowe* **Sv** *ternär fission*

Th ↦ *thorium*

theory of elasticity *mat* • mathematical study of how solid objects deform and become internally stressed due to prescribed loading conditions; *t.o.e.* is a branch of continuum mechanics $\mapsto$ strength of materials **D** *Elastizitätstheorie* **F** *théori de l'élasticité* **Pl** *teoria sprężystości* **Sv** *elasticitetsteori*

thermal column *rcs* • moderator body located in, or adjacent to, a nuclear reactor and in which nearly all neutrons are slowed down to the thermal neutron energy $\mapsto$ moderator; thermal neutron energy **D** *thermische Säule* **F** *colonne thermique* **Pl** *kolumna termiczna* **Sv** *termisk kolonn*

thermal conductivity *th* • (denoted λ or k, defined for a given material:) measure of a material's ability to conduct heat, defined through Fourier's law of heat conduction; for temperature-dependent *t.c.* it is convenient to introduce a conductivity integral defined as $I(T) = \int_{T_0}^{T} \lambda(T')dT'$, where T_0 is any reference temperature; in particular, for a nuclear fuel, a *conductivity integral to melt (CIM)* is defined as $CIM = \int_{T_0}^{T_m} \lambda(T')dT'$, where T_m is the fuel melting temperature $\mapsto$ heat conduction; Kirchhoff's transformation; Fourier's law of heat conduction **D** *Wärmeleitfähigkeit* **F** *conductivite thermique* **Pl** *przewodność cieplna* **Sv** *termisk konduktivitet*

thermal cross section *xr* • cross section for processes caused by thermal neutrons; because thermal neutrons have different energy distributions at different times (such as resulting from different temperatures) the term has no precise meaning, which is why often a cross section for neutrons of the velocity 2200 m/s is given instead $\mapsto$ cross section; thermal neutrons **D** *thermischer Wirkungsquerschnitt* **F** *section efficace thermique* **Pl** *termiczny przekrój czynny* **Sv** *termiskt tvärsnitt*

thermal diffusion process *nf* • isotope separation process based on the fact that, in a fluid with a temperature gradient, the heavy molecules have a higher concentration in the cold region $\mapsto$ isotope separation **D** *Thermodiffusionsprozeß* **F** *diffusion thermique; thermodiffusion* **Pl** *proces termodyfuzyjny* **Sv** *termisk diffusionsprocess*

thermal disadvantage factor $\mapsto$ *disadvantage factor*

thermal energy *th* • (for a thermodynamic system:) collective energy of relative motions of particles against the center of mass of the system, including the energy needed for phase change (e.g., from liquid to vapour) of the substance in the system $\mapsto$ energy; heat flux; heat transfer **D** *thermische Energie* **F** *énergie thermique* **Pl** *energia termiczna; energia cieplna* **Sv** *termisk energi; värmeenergi*

thermal fatigue *mat* • material damage due to persistent stress variations caused by temperature variations $\mapsto$ material fatigue **D** - **F** - **Pl** *zmęczenie cieplne (materiału)* **Sv** *termiskutmattning*

thermal fission *nap* • nuclear fission caused by thermal neutrons $\mapsto$ nuclear fission; thermal neutrons **D** *thermische Spaltung* **F** *fission thermique* **Pl** *rozszczepienie termiczne* **Sv** *termisk fission*

thermal-fission factor *rph* • (denoted η_T:) neutron yield per absorption for fissions caused by thermal neutrons, found as

$$\eta_T = \frac{\int_T \nu(E)\Sigma_f(E)\varphi(E)dE}{\int_T \nu(E)\Sigma_a(E)\varphi(E)dE},$$

where $\Sigma_f(E)$ and $\Sigma_a(E)$ are energy-dependent macroscopic cross sections of fuel material for fission and absorption, respectively, $\nu(E)$ is the energy-dependent neutron yield per fission, and $\varphi(E)$ is the thermal neutron flux density $\mapsto$ neutron yield per absorption; neutron yield per fission; thermal neutrons **D** *thermischer Spaltungsfaktor* **F** *facteur* *de fission thermique* **Pl** *rozszczepienie ter-miczne* **Sv** *termisk fissionsfaktor*

thermal growth *mat* ● (for a fuel rod in a reactor core:) length increase of the fuel rod subject to repeating temperature changes, e.g., during increase and decrease of the reactor core power ↦ fuel rod   **D** *thermisches Anschwellen* **F** *croissance thermique* **Pl** *wydłużenie cieplne* **Sv** *termisk tillväxt*

thermalization *rph* ● final phase of the slowing-down process of neutrons wherein the neutrons are approaching a thermal equilibrium with the surrounding material ↦ thermal neutrons   **D** *Thermalisierung* **F** *thermalisation* **Pl** *termalizacja* **Sv** *termalisering*

thermal neutron energy *rph* ● (denoted E, expressed in eV:) kinetic energy of thermal neutrons; assuming the Maxwellian distribution of the *t.n.e.*, the most probable neutron energy and the corresponding speed in a system with temperature $T = 293$ K are $E = 0.025$ eV and $v = 2200$ m/s, respectively ↦ neutron energy distribution; Maxwell's distribution law; thermal neutrons   **D** *thermische Neutronenenergie* **F** *énergie des neutrons thermiques* **Pl** *energia neutronów termicznych* **Sv** *termisk neutronenergi*

thermal neutrons *rph* ● neutrons completely or nearly completely in thermal equilibrium with the system through which they are diffusing, i.e., on average they gain as much energy as they lose due to scattering ↦ thermalization; thermal neutron energy; scattering   **D** *thermische Neutronen* **F** *neutrons thermiques* **Pl** *neutrony termiczne* **Sv** *termiska neutroner*

thermal power *roc* ● (in a nuclear reactor, usually expressed in MWt:) total heat produced in the reactor core per unit time ↦ heat   **D** *Wärmeleistung*

F *puissance thermique* **Pl** *moc cieplna* **Sv** *termisk effekt*

thermal radiation *bph* ● emission of electromagnetic radiation of wavelength primarily in the range 0.1 to 10 microns, from all matter that has a temperature greater than absolute zero; it represents the conversion of thermal energy into electromagnetic energy ↦ thermal energy; Stefan-Boltzmann law   **D** *Wärmestrahlung* **F** *rayonnement thermique* **Pl** *promieniowanie cieplne; promieniowanie termiczne* **Sv** *värmestrålning*

thermal reactor *rty* ● nuclear reactor in which nuclear fissions are mainly caused by thermal neutrons ↦ nuclear fission; thermal neutrons   **D** *thermischer Reaktor* **F** *réacteur à neutrons thermiques* **Pl** *reaktor termiczny* **Sv** *termisk reaktor*

thermal shield *rcs* ● radiation shield with high density and good thermal conductivity with the task of reducing the heat generation from ionizing radiation in surrounding regions and also the heat transport to these regions ↦ shield; ionizing radiation; heat transfer; heat conduction   **D** *thermische Abschirmung; thermischer Schild* **F** *bouclier thermique* **Pl** *osłona termiczna* **Sv** *termisk strålskärm; termisk skärm*

thermal spike *mat* ● spike in a crystal in which the transfer of kinetic energy from a primary knocked-on atom or another directly ionizing particle to the lattice atoms can be interpreted as a local and temporary heating ↦ spike; knocked-on atom; displacement spike   **D** *thermische Störzone* **F** *pointe thermique (dégât par rayonnements)* **Pl** *szczyt cieplny; szczyt temperaturowy* **Sv** *upphettningszon; smältzon*

thermal utilization factor *rph* ● (denoted f:) probability that a thermal neutron absorbed in a nuclear re-actor is absorbed in the fuel or in a specified fissile nuclide; for a homogeneous reactor the *t.u.f.* can be found as $f = \Sigma_{a,F}/\Sigma_a$, where $\Sigma_{a,F}$ and Σ_a are the macroscopic absorption cross sections for the fuel alone and for all materials in the core combined, respectively ↦ thermal neutrons; four-factor formula; infinite multiplication factor; fissile; nuclide **D** *thermische Nutzung* **F** *facteur d'utilisation thermique* **Pl** *współczynnik wykorzystania neutronów termicznych* **Sv** *termisk nyttofaktor*

thermodynamic equilibrium quality ↦*steam quality*

thermodynamic temperature ↦*absolute temperature*

thermoluminescence dosimeter *rdp* • dosimeter whose principle of operation is based on the thermoluminescence of a material such as, e.g., calcium fluoride or lithium fluoride ↦ dosimeter **D** *Thermolumineszenzdosimeter* **F** *dosimètre thermoluminescent* **Pl** *dawkomierz termoluminescencyjny* **Sv** *termoluminescensdosimeter*

thermonuclear reaction *bph* • nuclear reaction achieved by applying high temperature to the reacting mixture; the term generally refers to the fusion reaction ↦ nuclear fusion **D** *thermonukleare Reaktion* **F** *réaction thermonucléaire* **Pl** *reakcja termojądrowa* **Sv** *termonukleär reaktion*

thimble *rcs* • tube closed on one side containing control elements, experimental devices, or other items that are inserted in a nuclear reactor ↦ control element **D** *Fingerhutrohr* **F** *chaussette* **Pl** *tuleja rurowa* **Sv** *rörficka; provtub; tub*

thorium *mat* • chemical element denoted Th, with atomic number $Z = 90$, relative atomic mass $A_r =$ 232.0381, density 11.72 g/cm^3, melting point 1750 °C, boiling point 4788 °C, crustal average abundance 69.6 mg/kg and ocean abundance 1×10^{-6} mg/L, with only isotope ^{232}Th existing on Earth; *t.* is a fertile material that can be transmuted into fissile nuclide ^{233}U by a neutron capture ↦ fertile material; denatured uranium; element **D** *Thorium* **F** *thorium* **Pl** *tor* **Sv** *torium*

Three Mile Island nuclear accident *rs* • nuclear accident that occurred on March 28, 1979, in unit 2 of the Three Mile Island Nuclear Power Plant (TMI-2) in Dauphin County, Pennsylvania, near Harrisburg; it was the most significant accident in U.S. commercial nuclear power; the accident was caused by a combination of equipment failure and the inability of plant operators to understand the reactor's condition at certain times during the event; a gradual loss of cooling water to the reactor's heat-producing core led to partial melting of the fuel rod cladding and the uranium fuel, and the release of a small amount of radioactive material; the accident caused no injuries or deaths; experts concluded that the amount of radiation released into the atmosphere was too small to result in discernible direct health effects to the population in the vicinity of the plant ↦ Fukushima Daiichi nuclear accident; Chernobyl nuclear accident **D** *Unfall von Three Mile Island* **F** *accident nucléaire de Three Mile Island* **Pl** *awaria jądrowa w Three Mile Island* **Sv** *Harrisburgolyckan*

threshold detector *mt* • detector whose mode of operation is based on a threshold reaction ↦ threshold reaction; detector (of radiation) **D** *Schwellwertdetek-*

tor *F détecteur à seuil* **Pl** *detektor progowy* **Sv** *tröskeldetektor*

threshold dose *rdp* • minimum absorbed dose producing a specified effect ↦ absorbed dose   **D** *Schwellenwertdosis* **F** *dose seuil* **Pl** *dawka progowa* **Sv** *tröskeldos*

threshold reaction *nap* • nuclear reaction that takes place only if the incident particle has energy at least equal to a certain threshold energy; sometimes the *t.r.* refers to a reaction that occurs with a low probability below a certain energy value and with a high probability above that value ↦ nuclear reaction   **D** *Schwellwertreaktion* **F** *réaction à seuil* **Pl** *reakcja progowa* **Sv** *tröskelreaktion*

throughput *nf* • amount of fresh nuclear fuel that, within a given time interval, is inserted into the core in replacement of the spent fuel ↦ fuel; spent fuel   **D** *Durchsatz* **F** *capacité de passage* **Pl** *przerób (paliwa)* **Sv** *bränsleomsättning*

Ti ↦*titanium*

tie rod ↦*supporting fuel rod*

time-constant range *roc* • power interval of a nuclear reactor within which the time constant of the reactor, rather than its power, is the most important parameter for the reactor control ↦ power range   **D** *Zeitkonstantenbereich; Periodenbereich* **F** *domaine de divergence* **Pl** *zakres stałej czasowej (reaktora)* **Sv** *tidkonstantområde*

TIP system ↦*travelling in-core probe system*

tissue equivalent *rdp* • (of material:) whose scattering and absorption properties for a particular radiation types are the same as for a certain biological tissue ↦ scattering; absorption   **D** *gewebeäquivalent* **F** *équivalent au tissu*

Pl *równoważnik tkankowy* **Sv** *vävnadsekvivalent*

tissue equivalent ionization chamber *rdp* • ionization chamber in which the wall, electrode material, and gas are selected so that the generated ionization is a measure of the absorbed dose obtained in a biological tissue ↦ ionization chamber; absorbed dose   **D** *gewebeäquivalente Ionisationskammer* **F** *chambre d'ionisation équivalente au tissu; chambre d'ionisation équivalente au tissu biologique* **Pl** *komora jonizacyjna równoważna tkance biologicznej* **Sv** *vävnadsekvivalent jonkammare*

titanium *mat* • chemical element denoted Ti, with atomic number $Z=22$, relative atomic mass $A_r=47.867$, density 4.5 g/cm^3, melting point 1668 °C, boiling point 3287 °C, crustal average abundance 5650 mg/kg and ocean abundance 0.001 mg/L ↦ element   **D** *Titan* **F** *titane* **Pl** *tytan* **Sv** *titan*

toll enrichment *nf* • enrichment of a nuclear fuel provided against a payment ↦ enrichment[1]   **D** *Auftragsanreicherung* **F** *enrichissement à façon* **Pl** *odpłatne wzbogacenie paliwa* **Sv** *legoanrikning*

top guide ↦*core grid*

top nozzle *nf* • plate in the upper region of the PWR core supporting fuel assemblies; for the BWR core this part is called the *top tie plate* ↦ PWR; BWR; reactor core; bottom nozzle   **D** *obere Gitterplatte* **F** *plaque-support supérieure* **Pl** *płyta górna* **Sv** *topplatta*

top tie plate ↦*top nozzle*

total cross section *xr* • (for a particular substance and certain radiation:) sum of cross sections for all reaction types ↦ cross section; capture cross section; neutron absorption cross section; scattering cross section   **D** *totaler Wirkungsquerschnitt* **F** *section efficace to-*

tale **Pl** *całkowity przekrój czynny* **Sv** *totalt tvärsnitt*

TRACE *th* • (acronym for TRAC/RELAP Advanced Computational Engine:) modernized thermalhydraulic code designed to consolidate and extend the capabilities of USNRC's 3 legacy safety codes: TRAC-P, TRAC-B and RELAP; the *T.* code is able to analyse large- and small-break LOCA scenarios and system transients in both pressurized and boiling water reactors; a capability exists to model thermalhydraulic phenomena using both one-dimensional and three-dimensional components ↦ RELAP; SNAP **D** *TRACE* **F** *TRACE* **Pl** *TRACE* **Sv** *TRACE*

training reactor *rty* • nuclear reactor used mainly for education and training in the reactor operation area ↦ nuclear reactor **D** *Ausbildungsreaktor* **F** *réacteur d'entraînement* **Pl** *reaktor szkoleniowy* **Sv** *övningsreaktor*

transfer function *roc* • mathematical expression that relates the output value of a control system to its input value; for a nuclear reactor, the transfer function expresses, e.g., the relationship between the reactor power and the reactivity ↦ reactivity **D** *Übergangsfunktion* **F** *fonction de transfert* **Pl** *transformata* **Sv** *överföringsfunktion*

transgranular stress-corrosion cracking *mat* • stress-corrosion cracking across grains of a material ↦ stress-corrosion cracking; intergranular stress corrosion cracking **D** *transgranulare Spannungsrißkorrosion* **F** *fissuration transgranulaire sous contrainte* **Pl** *śródkrystaliczne pękanie wskutek korozji naprężeniowej* **Sv** *transkristallin spänningskorrosionssprickning*

transient *rs* • (during reactor operation:) event, other than a loss-of-coolant accident, leading to imbalance between supplied and removed heat in the reactor core; most transients are managed by the reactor control system without a need to interrupt the reactor operation; during some transients, a reactor trip is initiated to prevent overheating of the reactor core ↦ loss-of-coolant accident; reactor trip **D** *Transiente* **F** *transitoire* **Pl** *stan przejściowy* **Sv** *transient*

transmutation *bph* • (in physics:) changing of one chemical element or isotope into another one by radioactive decay, nuclear bombardment, or similar processes ↦ element; isotope; radioactive decay; conversion **D** *Transmutation* **F** *transmutation* **Pl** *transmutacja* **Sv** *transmutation*

transport cross section *xr* • (also called *macroscopic transport cross section*:) quantity defined as $\Sigma_{tr} = \Sigma_t - \overline{\mu}_0 \Sigma_s$, where Σ_t - total macroscopic cross section, $\overline{\mu}_0$ - average scattering angle cosine, and Σ_s - macroscopic scattering cross section ↦ macroscopic cross section; average scattering angle cosine; transport mean free path; total cross section; scattering cross section **D** *Transportquerschnitt* **F** *section efficace de transport* **Pl** *przekrój czynny na transport neutronów* **Sv** *transporttvärsnitt*

transport equation *rph* • linear equation, similar to the Boltzmann equation in the kinetic theory of gases, used in transport theory to describe the space and time dependence of the differential particle flux density ↦ transport theory; particle flux density **D** *Transportgleichung* **F** *équation du transport* **Pl** *równanie transportu* **Sv** *transportekvation*

transport index *rdp* • (usually expressed in the older unit mrem/h:) number assigned to a package to indicate its degree of danger; the *t.i.* is used to ensure radiation safety and protection against criticality during transport $\mapsto$ radiation; criticality; rem **D** *Transportindex* **F** *indice de transport* **Pl** *wskaźnik transportu* **Sv** *transportindex*

transport mean free path *rph* • (denoted λ_{tr}:) mean free path, in particular for neutrons, modified with respect to the anisotropy in the scattering; the *t.m.f.p.* is equal to the inverse of the macroscopic transport cross section: $\lambda_{tr} = 1/\Sigma_{tr}$ $\mapsto$ mean free path; scattering; transport cross section **D** *Transportweglänge* **F** *libre parcours moyen de transport* **Pl** *średnia droga swobodna transportu* **Sv** *transportväglängd*

transport theory *rph* • theory for neutrons and gamma radiation propagation in a medium, based on the transport equation $\mapsto$ transport equation **D** *Transporttheorie* **F** *théorie du transport* **Pl** *teoria transportu* **Sv** *transportteori*

transuranic element *mat* • all elements with atomic number $Z > 92$; essentially all the *t.e.* that appear in the nuclear waste are actinides $\mapsto$ actinide; transuranic waste **D** *Transuran; Transuranelement* **F** *transuranien* **Pl** *transuranowiec* **Sv** *transuran*

transuranic waste *wst* • (also called *TRU waste*:) radioactive waste containing transuranic elements above prescribed limits $\mapsto$ transuranic element **D** *Transuranabfall* **F** *déchets renfermant des éléments transuraniens* **Pl** *odpad transuranowy* **Sv** *transuranavfall*

travelling in-core probe system *mt* • (abbreviated *TIP system*:) system of movable neutron detectors used to determine the actual power distribution in a nuclear core and to calibrate the stationary detectors $\mapsto$ collectron; local power range monitor **D** *fahrbare Meßkammer* **F** *instrumentation interne du coeur; système RIC* **Pl** *system ruchomych detektorów w rdzeniu* **Sv** *TIP-system*

TRISO *nf* • (acronym for *Tri*structural-*Iso*tropic:) fuel type consisting of a fuel kernel composed of UOx in the center, coated with four layers of three isotropic materials; the four layers (1 - porous buffer layer made of carbon, 2 - dense inner layer of pyrolytic carbon, 3 - ceramic layer of silicon carbide, 4 - dense outer layer of pyrolytic carbon) retain fission products at elevated temperatures; the design of fuel prevents cracks due to the stresses up to 1600 °C $\mapsto$ coated particle; HTGR **D** *TRISO* **F** *TRISO* **Pl** *TRISO* **Sv** *TRISO*

tritium *mat* • (denoted ^3_1H, T or *t*:) radioactive isotope of hydrogen with relative atomic mass $A_r = 3.016\,049\,2$, beta-decaying into ^3He and having half-life 12.33 y; *t.* can be used as the fusion fuel in the most promising fusion reaction on Earth: $d + t \to \alpha + n + 17.59$ MeV; *t.* is extraordinarily rare on Earth, but it can be manufactured in nuclear reactors by reacting neutrons with lithium $\mapsto$ hydrogen; nuclear fusion; lithium **D** *Tritium* **F** *tritium* **Pl** *tryt* **Sv** *tritium*

TRU waste $\mapsto$ *transuranic waste*

tsunami *rs* • (also known as a *seismic sea wave*:) series of waves in a water body caused by the displacement of a large volume of water, generally in an ocean or a large lake, caused by earthquakes, volcanic eruptions and other underwater explosions, landslides, glacier calvings, meteorite impacts and other disturbances above or below water $\mapsto$ earthquake; Fukushima Daiichi nuclear accident **D** *Tsunami* **F** *tsunami* **Pl** *tsunami* **Sv** *tsunami*

turbine trip *roc* • interrupting the

steam supply to the turbine ↦ transient **D** *Turbinenauslösung* **F** *déclenchement de la turbine* **Pl** *odłączenie turbiny* **Sv** *turbinutlösning; turbinsnabbstängning*

turbogenerator *rcs* • electricity generator driven by a steam turbine or gas turbine ↦ reactor unit **D** *Turbogenerator* **F** *turbo-alternateur* **Pl** *turbogenerator; prądnica turbinowa* **Sv** *turbogenerator*

turbulence *th* • (in fluid dynamics:) property of fluid motion characterized by chaotic changes in pressure and flow velocity ↦ turbulent flow **D** *Turbulenz* **F** *turbulence* **Pl** *turbulencja; burzliwość* **Sv** *turbulens*

turbulent flow *th* • fluid flow with turbulent behaviour, typically occurring in pipes when the Reynolds number is greater than 2300 ↦ turbulence; Reynolds number **D** *turbulente Strömung* **F** *écoulement turbulente* **Pl** *przepływ turbulentny; przepływ burzliwy* **Sv** *turbulent strömning*

turn-around time *nf* • (for fuel reprocessing plant:) downtime for cleaning and preparation between various operation periods ↦ fuel reprocessing; shutdown **D** *Abschaltzeit zur Reinigung* **F** *temps de maintenance* **Pl** *czas konserwacji* **Sv** *omställningstid*

turning *roc* • maintaining the slow rotation of the turbine or generator to avoid thermal deformations ↦ turbogenerator **D** *Turning; Drehen* **F** *virage* **Pl** *obracanie (wirnika turbiny)* **Sv** *baxning*

two-group model *rph* • neutron transport model, where the neutrons are divided into two energy groups ↦ one-group model; multigroup model **D** *Zweigruppenmodell* **F** *modéle à deux groupes* **Pl** *model dwugrupowy* **Sv** *tvågruppsmodell*

two-group theory *rph* • theory of neutron transport based on the two-group model ↦ two-group model; multigroup model **D** *Zweigruppentheorie* **F** *théorie à deux groupes* **Pl** *teoria dwugrupowa* **Sv** *tvågruppssteori*

two-phase flow *th* • flow of a mixture of gas and liquid in a pipe or a cooling channel ↦ two-phase flow friction multiplier **D** *Zweiphasenströmung* **F** *écoulement diphasique* **Pl** *przepływ dwufazowy* **Sv** *tvåfasströmning*

two-phase flow friction multiplier *th* • (usually denoted ϕ_{rf}^2:) ratio defined as $\phi_{rf}^2 \equiv \left(\frac{dp}{dz}\right)_{2\phi} / \left(\frac{dp}{dz}\right)_{rf}$, where $\left(\frac{dp}{dz}\right)_{2\phi}$ is the friction pressure loss gradient in two-phase flow and $\left(\frac{dp}{dz}\right)_{rf}$ is the friction pressure loss gradient in a reference single-phase flow ↦ two-phase flow; Fanning friction factor; Darcy friction factor **D** - **F** *multiplicateur diphasique* **Pl** *mnożnik strat ciśnienia w przepływie dwufazowym* **Sv** *tvåfasmultiplikator*

two-phase flow pattern *th* • particular phase distribution in two-phase flow, depending on flow orientation, channel geometry and mass flow rates of liquid and gas phases ↦ two-phase flow **D** *Phasenverteilungszustände* **F** *configuration d'un écoulement diphasique* **Pl** *struktura przepływu dwufazowego* **Sv** *strömningstyp vid tvåfasströmning*

Uu

u ↦ *atomic mass unit*

U ↦ *uranium*

ultimate storage *wst* • installation for disposal of radioactive wastes that will not be further reprocessed or recycled ↦ radioactive waste **D** *Endlager* **F** *dépôt de stockage définitif* **Pl** *składowisko ostateczne* **Sv** *slutförvar*

ultimate waste disposal *wst* • disposal of nuclear wastes in the ultimate storage ↦ waste disposal; ultimate storage; interim storage; direct disposal **D** *Endlagerung²* **F** *destination finale; stockage définitif* **Pl** *ostateczne usuwanie odpadów* **Sv** *slutförvaring*

unclassified area *rdp* • area not classified from the radiation protection point of view; normally one must not store radioactive materials or conduct radiation work in the *u.a.* ↦ radiation protection; classified area; radioactive material **D** *nicht klassifizierter Bereich* **F** *zone non classée* **Pl** *strefa niesklasyfikowana* **Sv** *oklassat område*

underground ultimate storage *wst* • underground installation for disposal of radioactive wastes that will not be further reprocessed or recycled ↦ ultimate storage; geological repository **D** *unterirdisches Endlager* **F** *installation souterraine de stockage définitif* **Pl** *podziemne składowisko ostateczne* **Sv** *underjordsförvar*

undermoderated *rph* • (about multiplying system:) indicating that the volume ratio of moderator to fuel is less than the corresponding ratio of a reactor in which a certain physical parameter (such as the material buckling) has an extreme value ↦ multiplying; geometric buckling; material buckling; well-moderated; overmoderated **D** *untermoderiert* **F** *sous-modéré* **Pl** *niedomoderowany (układ mnożący)* **Sv** *undermodererad*

unified atomic mass unit ↦ *atomic mass unit*

universal gas constant *bph* • (denoted R:) physical constant defined as Avogadro's constant, N_A, multiplied by the Boltzmann constant, k_B; since both these constants have exact numerical values, the value of the *u.g.c.* is exactly $R = N_A k_B = 8.314\ 462\ 618\ 153\ 24$ J/(K·mol) ↦ ideal gas law; Avogadro's constant; Boltzmann constant **D** *universelle Gaskonstante* **F** *constante universelle des gaz parfaits* **Pl** *uniwersalna stała gazowa* **Sv** *allmänna gaskonstanten*

unsealed source *rd* • radiation source which is not a sealed source ↦ radiation source; sealed source; radioactive material **D** *offene Quelle* **F** *source non scellée* **Pl** *otwarte źródło promieniowania* **Sv** *öppen strålkälla*

unstable nuclide *rdy* • radioactive nuclide undergoing decay ↦ nuclide; stable nuclide **D** *instabiles Nuklid* **F** *nu-*

cléide instable **Pl** *nuklid nietrwały* **Sv** *instabil nuklid*

unusual event ↦*notification of unusual event*

upper drywell *rcs* • (for a boiling water reactor:) part of the drywell space in the containment, located above and around the reactor vessel ↦ containment; drywell; lower drywell **D** *oberer Primärkreisraum* **F** *enceinte sèche supérieure* **Pl** *górna komora pierwotna* **Sv** *övre primärutrymme*

upscattering *rph* • scattering in which the kinetic energy of a neutron increases ↦ scattering **D** *Aufwärtsstreuung* **F** *diffusion accélératrice* **Pl** *rozpraszanie przyspieszające* **Sv** *uppspridning*

uranium *mat* • chemical element denoted U, with atomic number $Z=92$, relative atomic mass $A_r=238.0289$, density 18.95 g/cm^3, melting point 1135 °C, boiling point 4131 °C, crustal average abundance 2.7 mg/kg and ocean abundance 0.0032 mg/L; most important are fissile isotopes ^{235}U (with isotopic abundance 0.7205%, α-decaying, $T_{1/2}=704$ My) and ^{233}U (obtainable from thorium, α-decaying, $T_{1/2}=0.158$ My) and fertile isotope ^{238}U (with isotopic abundance 99.2739%, α-decaying, $T_{1/2}=4470$ My) ↦ natural uranium; element **D** *Uran* **F** *uranium* **Pl** *uran* **Sv** *uran*

urban siting *gnt* • siting of a nuclear facility in or near a major population center ↦ siting criterion; site area **D** *stadtnaher Standort* **F** *établissement urbain* **Pl** *lokalizacja podmiejska* **Sv** *närförläggning*

urgent protective action planning zone ↦*emergency planning zone*

V v

V $\mapsto$ *vanadium*

value function *nf* • function used in the calculation of separative work in the isotope separation process, given by $V(x) = (1 - 2x)\ln\left(\frac{1-x}{x}\right)$, where x is the enrichment in weight fraction $\mapsto$ isotope separation; separative work **D** *Wertfunktion* **F** *fonction de valeur* **Pl** *funkcja wartości* **Sv** *värdefunktion*

vanadium *mat* • chemical element denoted V, with atomic number $Z=23$, relative atomic mass $A_r=50.9415$, density 6.0 g/cm³, melting point 1910 °C, boiling point 3407 °C, crustal average abundance 120 mg/kg and ocean abundance 0.0025 mg/L $\mapsto$ element **D** *Vanadium* **F** *vanadium* **Pl** *wanad* **Sv** *vanadin*

van der Waals equation of state *th* • modified ideal gas law, which takes into account non-zero volumes and mutual forces of gas particles; the reduced form of the equation is as follows:

$$\left(p_r + \frac{3}{v_r^2}\right)\left(v_r - \frac{1}{3}\right) = \frac{8}{3}T_r,$$

here $p_r = p/p_{cr}, v_r = v/v_{cr}, T_r = T/T_{cr}$ with f_r and f_{cr} representing reduced and critical variables respectively, and p - pressure, v - specific volume and T - absolute temperature $\mapsto$ ideal gas law **D** *Van-der-Waals-Gleichung* **F** *équation d'état de van der Waals* **Pl** *równanie stanu van der Waalsa* **Sv** *van der Waals tillståndsekvation*

very high temperature reactor *rty* • (abbreviated *VHTR:*) generation IV reactor concept that uses a graphite-moderated nuclear reactor with a once-through uranium fuel cycle; gas coolant can have an outlet temperature of 1000 °C; the reactor core can be either a prismatic block or a pebble-bed core; the high outlet temperatures enable applications such as process heat or hydrogen production via the thermochemical sulfur–iodine cycle $\mapsto$ generation IV reactor; high-temperature gas-cooled reactor; graphite reactor **D** *Höchsttemperaturreaktor* **F** *réacteur à très haute température* **Pl** *bardzo wysokotemperaturowy reaktor chłodzony gazem* **Sv** *mycket hög temperatur reaktor*

VHTR $\mapsto$ *very high temperature reactor*

virgin neutron *rph* • neutron that has not yet undergone its first collision after being sent from a source of any kind $\mapsto$ neutron **D** *jungfräuliches Neutron* **F** *neutron vierge* **Pl** *neutron pierwotny; neutron dziewiczy* **Sv** *okolliderad neutron; jungfruneutron*

vitrification *wst* • solidification of radioactive waste by calcination and fusing the calcine to a glass mass $\mapsto$ radioactive waste; calcination; waste-disposal plant **D** *Verglasung* **F** *vitrification* **Pl** *witryfikacja; zeszklenie* **Sv** *förglasning; vitrifiering*

void *rph* • effectively empty space

173

from the point of view of neutron balance, which arises in coolant of a nuclear reactor core by coolant evaporation during boiling ↦ void coefficient of reactivity; boiling reactor **D** *Hohlraum; Dampfvolumen* **F** *vide* **Pl** *próżnia* **Sv** *void*

void coefficient of reactivity *rph* • reactivity coefficient with respect to the void fraction in a certain part of the nuclear reactor core ↦ void fraction; reactivity coefficient **D** *Blasenkoeffizient der Reaktivität* **F** *coefficient de vide* **Pl** *próżniowy współczynnik reaktywności* **Sv** *voidkoefficient; ångblåskoefficient*

void fraction *rph* • volume fraction of the empty space due to formation of steam in a reactor core ↦ void coefficient of reactivity; void **D** *Hohlraumanteil; Dampfgehalt* **F** *fraction de vide* **Pl** *współczynnik próżni* **Sv** *voidhalt*

volatility process *nf* • chemical separation process based on different degrees of volatility of the constituents in a mixture; the process is primarily used to purify uranium and to separate it from the plutonium during fuel reprocessing and is based on the high volatility of uranium hexafluoride ↦ fuel reprocessing **D** *Verflüchtigungsverfahren* **F** *purification par distillation fractionnée* **Pl** *proces lotnościowy* **Sv** *förångningsprocess*

volume reduction factor *nf* • (abbreviated *VRF*:) ratio of the volume of radioactive waste before and after treatment ↦ radioactive waste **D** *Volumenverkleinerungsfaktor* **F** *facteur de concentration de volume* **Pl** *współczynnik redukcji objętości* **Sv** *volymreduktionsfaktor*

VRF ↦ *volume reduction factor*

Ww

W ↦watt

walkaway safety *rs* • (at nuclear installation:) feature meaning that safety of the nuclear installation is maintained for a long time, without human intervention or supervision required ↦ inherent safety; nuclear installation **D** - **F** - **Pl** *bezpieczeństwo samozachowawcze* **Sv** *självbevarande säkerhet*

waste conditioning *wst* • waste treatment aimed at meeting conditions for storage, transport or ultimate disposal ↦ waste treatment; ultimate waste disposal **D** *Abfallkonditionierung* **F** *conditionnement des déchets* **Pl** *kondycjonowanie odpadów* **Sv** *avfallskonditionering*

waste disposal *wst* • final taking care of, or getting rid of, nuclear waste ↦ nuclear waste **D** *Abfallbeseitigung* **F** *évacuation des déchets* **Pl** *usuwanie odpadów* **Sv** *avfallsdisponering*

waste-disposal plant *wst* • plant intended for treatment or disposal of radioactive waste ↦ radioactive waste; waste disposal **D** *Abfallbeseitigungsanlage* **F** *installation de traitement des déchets* **Pl** *zakład usuwania odpadów* **Sv** *avfallsanläggning*

waste form *wst* • physical or chemical form of radioactive waste, which may be liquid or solid ↦ radioactive waste; waste-disposal plant; waste disposal **D** *Abfallart* **F** *nature des déchets; forme de déchets* **Pl** *rodzaj odpadów* **Sv** *avfallsform*

waste management *wst* • measures taken with nuclear wastes, including their treatment, transportation and disposal ↦ waste disposal; waste treatment; waste conditioning **D** *Abfallbehandlung*[1] **F** *gestion des déchets* **Pl** *gospodarka odpadami* **Sv** *avfallshantering*

waste package *wst* • waste together with a container, constituting a unit for waste processing, transport, keeping or storage ↦ waste disposal; waste management **D** *Abfallbehälter* **F** *colis de déchets* **Pl** *pakunek z odpadami* **Sv** *avfallskolli*

waste plant ↦waste-disposal plant

waste treatment *wst* • separation, concentration, dilution, solidification or other transformation of radioactive waste ↦ waste disposal; waste management **D** *Abfallbehandlung*[2] **F** *traitement des déchets* **Pl** *obróbka odpadów* **Sv** *avfallsbehandling*

water-cross fuel assembly *nf* • fuel assembly with a central water cross; the *w.-c.f.a.* is intended for boiling water reactors ↦ fuel assembly **D** *Brennelementbündel mit Wasserkreutz* **F** *assemblage combustible avec canal d'eau central* **Pl** *kaseta paliwowa z krzyżowym kanałem wodnym* **Sv** *bränslepatron med vattenkors*

waterhammer *th* • pressure surge or wave caused when a fluid in motion (usually liquid but sometimes also gas) is suddenly forced to stop or change direction; the *w.* commonly occurs when a valve suddenly closes and a pressure wave propagates in the pipe ↦ isolation; isolation valve; transient **D** *Wasserhammer* **F** *coup de bélier* **Pl** *uderzenie hydrauliczne; uderzenie wodne* **Sv** *vattenslag*

waterlogging *nf* • ingress of water in a fuel element through a defect in the cladding ↦ fuel element; cladding **D** *Wassereinbruch* **F** *pénétration d'eau* **Pl** *penetracja wody* **Sv** *vatteninträngning*

water-moderated reactor *rty* • nuclear reactor moderated with water such as, e.g., a light-water reactor or a heavy-water reactor ↦ light-water reactor; heavy-water reactor; pressurized water reactor; boiling water reactor **D** *wassermoderierter Reaktor* **F** *réacteur modéré à l'eau* **Pl** *reaktor z moderatorem wodnym* **Sv** *vattenreaktor*

water monitor *rdp* • radiation monitor for water that contains a radioactive material ↦ radioactive material **D** *Wasserwarngerät* **F** *moniteur de contamination de l'eau* **Pl** *wskaźnik kontaminacji wody* **Sv** *vattenmonitor*

watt *bph* • (denoted W:) unit of power derived as one joule per second; frequently used multiples of the watt include $1kW = 10^3$ W, 1 MW $= 10^6$ W, 1 GW $= 10^9$ W and 1 TW $= 10^{12}$ W; in the electric power industry, megawatt electrical (MWe) is used for electric power, whereas megawatt thermal (MWt) is used for thermal power ↦ erg; thermal power **D** *Watt* **F** *watt* **Pl** *wat* **Sv** *watt*

Watt's distribution law *rph* • function describing distribution of the kinetic energy of prompt neutrons given as $X_p(E) = ae^{-E/b}\sinh\left(\sqrt{cE}\right)$, where E - neutron energy, $X_p(E)$ - neutron energy distribution such that $X_p(E)dE$ is equal to the neutron frac-

tion having energy between E and $E+dE$, and a, b, c are constants, which depend on the type of fission reaction; for fission of ^{235}U with thermal neutrons, $a = 0.5535, b = 1.0347, c = 1.6214$ ↦ Maxwell's distribution law; neutron energy distribution **D** *Wattverteilung* **F** *loi de distribution de Watt* **Pl** *rozkład Watta* **Sv** *Wattfördelning*

Weber number *th* • (denoted We:) dimensionless number in fluid mechanics that is often useful in analysing fluid flows where there is an interface between two different fluids, especially for multiphase flows with strongly curved surfaces; it represents a measure of the relative importance of the fluid's inertia compared to its surface tension; the *W.n.* is useful in analysing thin film flows and the formation of droplets and bubbles and is given as $We = \rho U^2 L/\sigma$, where ρ - density, U - velocity, L - characteristic length and σ - surface tension ↦ Reynolds number **D** *Weber-Zahl* **F** *nombre de Weber* **Pl** *liczba Webera* **Sv** *Webers tal*

Weisman correlation *th* • experimental correlation used to determine the Nusselt number as a function of the Reynolds and Prandtl numbers for turbulent heat transfer to non-metallic liquids and gases in fuel rod assemblies, given as: $Nu = CRe^{0.8}Pr^{1/3}$, where $C = 0.042s/d - 0.024$ with $1.1 \leq s/d \leq 1.3$ for a square lattice and $C = 0.026s/d - 0.006$ with $1.1 \leq s/d \leq 1.5$ for a triangular lattice, in which s is the lattice pitch and d is the rod diameter ↦ Dittus-Boelter correlation; Colburn correlation; Markoczy correlation; Nusselt number; pitch **D** *Weisman Gleichung* **F** *corrélation de Weisman* **Pl** *korelacja Weismana* **Sv** *Weismans korrelation*

well-moderated *rph* • (about multiplying system:) indicating that the volume ratio of moderator to fuel is such that the low-energy part of the neutron spectrum can be approximated with the Maxwell distribution and that most of the neutrons belong to this distribution ↦ multiplying; Maxwell's distribution law; undermoderated; overmoderated **D** *gutmoderiert* **F** *bien modéré* **Pl** *dobrze moderowany (układ mnożący)* **Sv** *välmodererad*

Westcott g-factor *rph* • (in Westcott model:) factor depending on the neutron temperature and indicating the cross section's deviation from the $1/v$-law in the thermal neutron region ↦ Westcott model **D** *Westcott-Faktor g* **F** *facteur g de Westcott* **Pl** *współczynnik g Wescotta* **Sv** *Westcotts g-faktor*

Westcott model *rph* • model for calculating the effective thermal cross section based on the assumption that a neutron flux density per energy interval has a Maxwell distribution for thermal neutrons and is inversely proportional to the neutron energy for the epithermal neutrons; according to the *W.m.* the effective thermal cross section is given as $\sigma_{eff} = \sigma_0(g + rs)$, where σ_0 is the cross section at neutron speed 2200 m/s, g is the Westcott g-factor, r is the Westcott r-factor and s is the Westcott s-factor ↦ Westcott g-factor; Westcott r-factor; Westcott s-factor **D** *Westcott-Modell; Westcott-Schreibweise* **F** *modèle de Westcott* **Pl** *model Westcotta* **Sv** *Westcotts modell*

Westcott r-factor *rph* • (in Westcott model:) factor depending on neutron temperature that is a measure of the fraction of epithermal neutrons, for example, in a nuclear reactor ↦ Westcott model **D** *Westcott-Faktor r* **F** *facteur r de Westcott* **Pl** *współczynnik r Westcotta* **Sv** *Westcotts r-faktor*

Westcott s-factor *rph* • (in Westcott model:) factor depending on neutron temperature that is a measure of the resonance integral with $1/v$-part subtracted ↦ Westcott model **D** *Westcott-Faktor s* **F** *facteur s de Westcott* **Pl** *współczynnik s Westcotta* **Sv** *Westcotts s-faktor*

wet fraction *th* • ratio of the liquid mass flow rate to the total mass flow rate in a two-phase flow equal to $1-x$, where x is the steam quality ↦ two-phase flow; steam quality **D** *Feuchtigkeitsgrad* **F** - **Pl** *stopień zawilgocenia* **Sv** *fukthalt*

wetwell *rcs* • space in the reactor containment that contains cold water or ice for condensation of vapour from the pressure relief system or the blow-down system ↦ containment; pressure relief system; blowdown system; drywell **D** *Kondensationsraum* **F** *volume de condensation* **Pl** *komora wtórna; komora kondensacyjna* **Sv** *sekundärutrymme; kondensationsutrymme*

whole-body counter *rdp* • equipment to measure the total gamma radiation (including the bremsstrahlung) emitted from a human body, wherein one or more heavily shielded detectors are used against the radiation from the environment ↦ gamma radiation; bremsstrahlung **D** *Ganzkörperzähler* **F** *anthroporadiamètre* **Pl** *licznik do pomiaru promieniowania całego ciała* **Sv** *helkroppsmätare*

wide range monitor *roc* • power range monitor covering two or all three operating ranges: the source range monitor, the intermediate range monitor and the power range monitor ↦ operating range; source range monitor; intermediate range monitor; power range monitor **D** *Breitkanalmonitor* **F** - **Pl** *monitor szerokozakresowy* **Sv** *bredeffektkanal*

Wien's displacement law *bph* •

physical law stating that the spectral radiance of the black body radiation per unit wavelength peaks at the λ_{max} given by $\lambda_{max} = b/T$, where T is the absolute temperature in kelvin, and b is a constant of proportionality called Wien's displacement constant equal to $2.897\ 772\ 9(17) \times 10^{-3}$ m·K $\mapsto$ Stefan-Boltzmann law **D** *Wiensches Verschiebungsgesetz* **F** *loi du déplacement de Wien* **Pl** *prawo przesunięć Wiena* **Sv** *Wiens förskjutningslag*

Wigner effect *mat* • change in physical properties of graphite occurring during reactor operation due to displacements of the lattice atoms caused by high-energy neutrons or other high-energy particles $\mapsto$ Wigner energy **D** *Wigner-Effekt* **F** *effet Wigner* **Pl** *efekt Wignera* **Sv** *Wigner-effekt*

Wigner energy *mat* • energy stored in graphite by the Wigner effect $\mapsto$ Wigner effect **D** *Wigner-Energie* **F** *énergie Wigner* **Pl** *energia Wignera* **Sv** *Wigner-energi*

Wigner-Wilkins method *rph* • method of calculation of the thermal neutron spectrum in hydrogen-containing materials, using the free-gas model for hydrogen nuclei $\mapsto$ thermal neutron; free-gas model **D** *Wigner-Wilkins-Methode* **F** *méthode de Wigner-Wilkins* **Pl** *metoda Wignera-Wilkinsa* **Sv** *Wigner-Wilkins metod*

Xx

Xe $\mapsto$ *xenon*

xenon *mat* • chemical element denoted Xe, with atomic number $Z=54$, relative atomic mass $A_r=131.29$, density 2.953 g/cm^3, melting point -111.75 °C, boiling point -108.04 °C, crustal average abundance 3×10^{-5} mg/kg and ocean abundance 5×10^{-5} mg/L; isotope ^{135}Xe is the most important nuclear poison with the microscopic cross section for absorption $\sigma_a = 2.75\times10^6$ b $\mapsto$ element; neutron absorption cross section; nuclear poison; xenon poisoning **D** *Xenon* **F** *xénon* **Pl** *ksenon* **Sv** *xenon*

xenon effect *rph* • reactivity reduction due to xenon poisoning $\mapsto$ reactivity; xenon; xenon poisoning **D** *Xenoneffekt* **F** *effet xénon* **Pl** *zjawisko ksenonowe* **Sv** *xenonverkan*

xenon equilibrium *rph* • state in a nuclear reactor when a balance between generation and disappearance of the nuclear poison ^{135}Xe is achieved; ^{135}Xe is directly or indirectly formed from the fission process and disappears due to neutron capture and radioactive decay $\mapsto$ nuclear poison; fission; radioactive decay **D** *Xenongleichgewicht* **F** *équilibre xénon* **Pl** *równowaga ksenonowa* **Sv** *xenonjämvikt*

xenon instability *roc* • local power oscillation in a large nuclear reactor caused by the dependence of xenon poisoning on the thermal neutron flux density $\mapsto$ xenon poisoning; neutron flux density **D** *Xenoninstabilität; Xenonschwingung* **F** *instabilité xénon* **Pl** *oscylacja ksenonowa* **Sv** *xenonsvängning*

xenon override *rph* • part of a nuclear reactor excess reactivity that is intended to allow the reactor to start even when the xenon effect reached its maximum value after shutdown $\mapsto$ excess reactivity; xenon effect; shutdown **D** *Xenonreaktivitätsreserve* **F** *surréactivité anti-xénon* **Pl** *kompensacja ksenonowa* **Sv** *xenonkompensation*

xenon poisoning *rph* • forming of fission product ^{135}Xe, which is a nuclear poison, in a thermal reactor $\mapsto$ nuclear poison **D** *Xenonvergiftung* **F** *empoisonnement xénon* **Pl** *zatrucie ksenonem* **Sv** *xenonförgiftning*

xenon transient *roc* • transient of reactivity due to xenon poisoning caused by a local or global change in the nuclear reactor power $\mapsto$ transient; reactivity; xenon poisoning **D** *Xenontransiente* **F** *transitoire xénon* **Pl** *stan przjściowy wywołany zatruciem ksenonem* **Sv** *xenontransient*

X radiation *rd* • electromagnetic radiation with wave length in a range from 10^{-11} m to 5×10^{-5} m, created due to deceleration of charged particles through matter or due to electron jumps from high-energy to low-energy electron shells in atoms of a substance bombarded with a

beam of electrons with energy from
1 to 500 keV ↦ gamma radiation;
electron *D Röntgenstrahlung F rayon-*
nement X Pl promieniowanie rentgenowskie;
promieniowanie X Sv röntgenstrålning
X-ray radiography *mt* • radiogra-

phy with X radiation ↦ radiography[1];
X radiation *D Röntgenographie F ra-*
diographie aux rayons X Pl radiografia
rentgenowska Sv röntgenradiografi

X rays ↦*X radiation*

Y y

yellow cake *nf* • intermediate product in the uranium fuel cycle obtained after calcination of U_3O_8 (uranium oxide), containing about 70–90% of U_3O_8; the *y.c.* is a form of uranium most commonly traded in commodity markets; further calcination and solvent extraction is used to refine the *y.c.* to essentially pure UO_3 ↦ fuel cycle; calcination; solvent extraction **D** *Urangelb* **F** *concentré uranifère* **Pl** *koncentrat rudy uranu* **Sv** *urankoncentrat*

yield strength ↦*strength of materials*

yield stress ↦*strength of materials*

Young's modulus of elasticity ↦*strength of materials*

Yvon's method ↦*double spherical harmonics method*

Zz

zero-energy reactor *rty* • nuclear reactor for use in such a low power range that no cooling system is needed ↦ power reactor   **D** *Nulleistungsreaktor* **F** *réacteur de puissance nulle* **Pl** *reaktor mocy zerowej* **Sv** *nolleffektsreaktor*

zero-power reactor ↦*zero-energy reactor*

zinc *mat* • chemical element denoted Zn, with atomic number $Z=30$, relative atomic mass $A_r=65.39$, density 7.134 g/cm^3, melting point 419.53 °C, boiling point 907 °C, crustal average abundance 70 mg/kg and ocean abundance 0.0049 mg/L; the isotope ^{64}Zn (with natural abundance 48.9%) can be activated and transmuted into radioactive isotope ^{65}Zn as a result of the (n,γ) reaction, with half-life 250d and maximum gamma-ray energy 1.11 MeV ↦ element; induced radioactivity; radioactive half-life   **D** *Zink* **F** *zinc* **Pl** *cynk* **Sv** *zink*

zircaloy *mat* • group of zirconium alloys used in nuclear reactors for fuel cladding and for other purposes; *z.* is mainly characterized by the low cross section for neutron absorption ↦ zir-conium; cladding; neutron absorption   **D** *Zirkaloy* **F** *zircaloy* **Pl** *zircaloy* **Sv** *zircaloy*

zirconium *mat* • chemical element denoted Zr, with atomic number $Z=40$, relative atomic mass $A_r=91.224$, density 6.52 g/cm^3, melting point 1855 °C, boiling point 4409 °C, crustal average abundance 165 mg/kg and ocean abundance 3×10^{-5} mg/L; *z.* is used as an alloy component in zircaloy ↦ element; zircaloy   **D** *Zirconium* **F** *zirconium* **Pl** *cyrkon* **Sv** *zirkonium*

Zn ↦*zinc*

zone for prompt area contamination measurement *rdp* • area with a radius of 50 kilometres around a nuclear power plant with predetermined measuring points for the rapid measurement at high emissions ↦ emergency planning zone; exclusion area   **D** *Zone mit sofortiger Messung der Flächenkontamination* **F** *zone concernée par des contrôles rapides de la contamination surfacique; zone pour mesure instantanée de la surface de contamination* **Pl** *strefa natychmiastowego pomiaru skażenia* **Sv** *indikeringszon*

Zr ↦*zirconium*

Index of German Entries

absolute Temperatur ↦ *absolute temperature*, 3

Absorber ↦ *absorber*, 3

Absorption ↦ *absorption*, 3

Absorptionsmittel ↦ *absorber*, 3

Abstandsgestell ↦ *bird cage*, 17

Abstandshalter ↦ *spacer*, 153

Abstand zur kritischen Heizflächenbelastung ↦ *dryout margin*, 49

Abstreifkonzentration einer Kaskade ↦ *cascade tails assay*, 25

ACR ↦ *ACR*, 4

Adjungierte der Neutronenflußdichte ↦ *adjoint flux*, 5

Akkumulatorsystem ↦ *accumulator system*, 4

Aktinide ↦ *actinide*, 4

Aktinium ↦ *actinium*, 4

aktive Prüfung ↦ *hot testing*, 76

Aktivierung ↦ *activation*, 5

Aktivierungsdetektor ↦ *activation detector*, 5

Aktivierungsfolie ↦ *activation foil*, 5

Aktivierungsprodukt ↦ *activation product*, 5

Aktivierungsquerschnitt ↦ *activation cross section*, 5

Aktivität ↦ *activity*, 5

Aktivitätskonzentration ↦ *activity concentration*, 5

ALARA ↦ *ALARA*, 6

Albedo ↦ *albedo*, 6

Albedodosimeter ↦ *albedo dosimeter*, 6

Alphakasten ↦ *alpha box*, 6

Alphastrahler ↦ *alpha emitter*, 7

Alphastrahlung ↦ *alpha radiation*, 7

Alphateilchen ↦ *alpha particle*, 7

Alphazerfall ↦ *alpha decay*, 6

Aluminium ↦ *aluminum*, 7

Americium ↦ *americium*, 7

Andrückhülle ↦ *collapsible cladding*, 29

Anfahrbrennelement ↦ *booster element*, 18

Anfahren ↦ *running up*, 144

Anfahren des Reaktors ↦ *reactor start-up*, 138

Anfahrneutronenquelle ↦ *start-up neutron source*, 156

Anfang des Brennstoffkreislaufs ↦ *front-end of the fuel cycle*, 66

anfängliches Konversionsverhältnis ↦ *initial conversion ratio*, 81

anfängliche Überschußreaktivität ↦ *built-in reactivity*, 21

angereicherter Brennstoff[1] ↦ *enriched fuel[1]*, 55

angereicherter Brennstoff[2] ↦ *enriched fuel[2]*, 55

angereichertes Material ↦ *enriched material*, 55

angereichertes Uran[1] ↦ *enriched uranium[1]*, 55

angereichertes Uran[2] ↦ *enriched uranium[2]*, 55

angewandte Strahlenphysik ↦ *radiological physics*, 134

Annäherung an den kritischen Zustand ↦ *approach to criticality*, 8

Anreicherung[1] ↦ *enrichment[1]*, 56

Anreicherung[2] ↦ *enrichment[2]*, 56

Anreicherungsfaktor ↦ *enrichment factor*, 56

Anreicherungsgrad ↦ *degree of enrichment*, 41

Anteil der prompten Neutronen ↦ *prompt neutron fraction*, 126

Anteil der verzögerten Neutronen ↦ *delayed neutron fraction*, 41

Index of French Entries

Index of Polish Entries

Index of Swedish Entries

282